EARTH ON FIRE

Brewing Plagues and Climate Chaos in Our Backyards

EARTH ON FIRE

Brewing Plagues and Climate Chaos in Our Backyards

Evaggelos Vallianatos

World Scientific

NEW JERSEY · LONDON · SINGAPORE · BEIJING · SHANGHAI · HONG KONG · TAIPEI · CHENNAI · TOKYO

Published by

World Scientific Publishing Co. Pte. Ltd.

5 Toh Tuck Link, Singapore 596224

USA office: 27 Warren Street, Suite 401-402, Hackensack, NJ 07601

UK office: 57 Shelton Street, Covent Garden, London WC2H 9HE

Library of Congress Control Number: 2025035017

British Library Cataloguing-in-Publication Data
A catalogue record for this book is available from the British Library.

The opinions expressed in this work are solely the author's own and don't reflect the opinions or beliefs of the publisher and its affiliates.

ISBN 978-981-98-0094-0 (hardcover)
ISBN 978-981-98-0151-0 (paperback)
ISBN 978-981-98-0095-7 (ebook for institutions)
ISBN 978-981-98-0096-4 (ebook for individuals)

For any available supplementary material, please visit
https://www.worldscientific.com/worldscibooks/10.1142/14044#t=suppl

Desk Editors: Nambirajan Karuppiah/Zerlina Zhuang/Amanda Yun

Typeset by Stallion Press
Email: enquiries@stallionpress.com

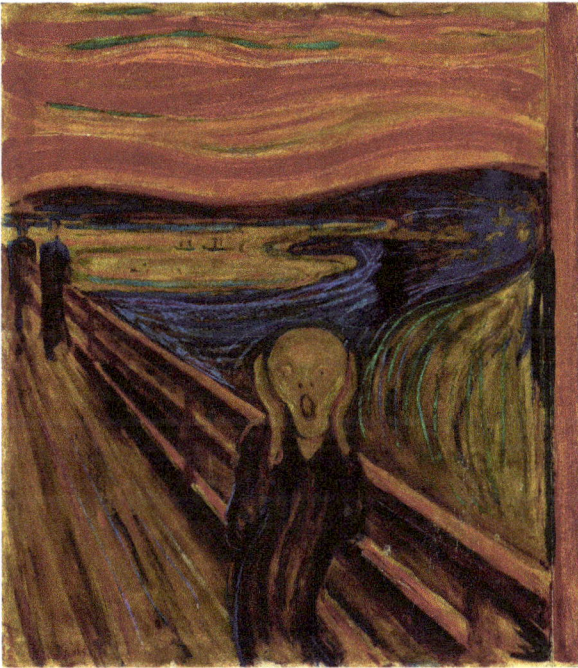

Edvard Munch, The Scream, 1893. National Gallery of Norway. Public Domain.

The painting reflected Munch's experience of existential dread while out on a walk. Hence, we could also interpret it as Munch's horror for the violence against nature. It is prophetic of the anthropogenic cosmic tragedy of rising temperature and irreversible damage done to the Earth, humans and civilization. The man/phantom engulfed in terror while crossing a bridge to nowhere with an ominous sky, and a bright orange background reminding us of those wild fires raging in the US.

About the Author

Evaggelos Vallianatos studied zoology and ancient and medieval Greek history at the University of Illinois. He earned a PhD in modern Greek and European history at the University of Wisconsin–Madison. He did postdoctoral studies in the history of science at Harvard University. He is a historian and ecopolitical theorist, former analyst of the ecological and social effects of industrialized farming at Capitol Hill, Congressional Office of Technology Assessment and office of Congressman Clarence Long of Maryland, as well as analyst of the risks of pesticides and the benefits of organic farming and biological and cultural alternatives to pesticides for the US Environmental Protection Agency. Evaggelos has taught at the American University, George Washington University, Humboldt State University, Bard College, University of New Orleans, University of Maryland, Pitzer College and California State Polytechnic University, Pomona. He authored hundreds of articles and several books, including *Poisson Spring* (2014), *The Antikythera Mechanism* (2021) and *Freedom* (2025).

Contents

Introduction

I was born and brought up in a Greek village. My father owned a few acres of land, divided up into small sections for the growing of wheat, barley, lentils, grape vines for wine and olive trees for olive oil. I worked with my family in the raising of our food. At the age of eighteen, I left the village for America where I arrived on August 5, 1961, a year before Rachel Carson published her insightful book, *Silent Spring*. I went to school straightaway and spent ten years studying biology and history. But when I graduated from the University of Wisconsin in 1972 with a doctorate in history, I had never heard of Rachel Carson or the US Environmental Protection Agency (EPA), the federal department brought into being in December 1970 in order to "protect" public health and nature from the toxins of industrialized and industrializing America.

It was my post-doctoral studies in the history of science at Harvard in the early 1970s that forced me out of my student innocence. My research and writing opened my eyes to the nature and politics of science, a powerful force shaping the anthropogenic impacts on nature and society. I did field research in Colombia in order to observe the so-called "miracles" of industrial agriculture, then known as the "green revolution." I discovered that, starting in the 1950s, this so-called green revolution had the support of the World Bank, the Ford Foundation, and the Rockefeller Foundation. The entire enterprise was based on large farmers owning big machines and using abundant quantities of deleterious chemical pesticides.

In addition, I noticed how American and international experts, in the name of corporate profits, worked hand in hand with Colombia's agrarian oligarchy. This American support of the richest and largest Colombian farmers undermined the land and culture of the small farm owners. The foreigners were certain that the mechanized and chemical agriculture, which they brought with them to Colombia, would boost the production of cash crops in the country's fertile valleys. They wanted the hard currency their cash crops would fetch in the markets of North America and Europe.

I described the "green revolution" in my 1976 book, *Fear in the Countryside: The Control of Agricultural Resources in the Poor Countries by Non-Peasant Elites*. The title tells the story. I sensed the fear Colombian small farmers felt for their lives and survival. The country was in the midst of civil and agrarian wars. A Colombian agronomist who drove me all over the country was telling me stories of the peasants trying for decades to maintain their freedom. Authorities advised us to avoid visiting certain provinces wrecked by violence and war. Both violence and war divided Colombians into two camps. One was the American-supported oligarchy and the other included the vast rural population.

By the time I got to Washington, DC, in December 1975, for my first job with the Congressional Office of Technology Assessment, I was well on my way to learning more and more about the practical and political lessons of scientific knowledge. Yet I stood by my values that the small family farmers' ancient methods have always been more tuned to the Earth than the industrial agricultural processes of North America and Europe; and that pesticides, heavy tractors, and scientists trained at Iowa State University, did not fit in the tropics. If anything, I used to tell my friends and colleagues in Congress that America needed agrarian or land reform so that farmers, land, science and technology would once again be brought in line with the values of democracy and the requirements of nature.

My protest went nowhere. The country was torn by the killings and aftermath of the Vietnam War. The celebration of Earth Day and the reading of Rachel Carson's *Silent Spring* were soothing to the disillusioned young college students who saw the idea of ecology as an alternative to the wholesale wrecking of society and nature going on in the jungles of Southeast Asia. In fact, EPA was set up in the same year as the first Earth Day, 1970. This was the work of the same Richard Nixon who spread the ferocity of the Vietnam War to peaceful Cambodia in secret during the Christmas days of 1970, only moments away from Nixon's proclamation giving birth to the EPA.

Meanwhile, my head was full of the lessons I learned during the writing of *Fear in the Countryside*. The forces in the United States exercising political influence on behalf of farm machinery, chemicals, and science were simply too large and powerful. Their impact in rural America was overwhelming to small family farmers. Agribusiness acted with liberty, taking over the land of the small farmers.[1] In January 1981, Secretary of Agriculture, Bob Bergland, issued a report in which he summed up the trends like increasing the size of farms and diminishing the number of small family farmers. "We thought – we hoped," Bergland said, "that if we helped the major commercial farmers, who provided most of our food and fiber (and exerted most of the political pressure), the benefits would filter down to the intermediate-sized and then the smallest producers. I was never convinced we were anywhere near the right track. We had symbols, slogans, and superficialities. We seldom had substance.... The success of our agriculture is true, but it is also true that, by 1978, about 7.7% of the households in America owned all the farm and ranch land. Of those households, 62,260 – the population of a medium-sized city – owned three of every ten acres. How did this come about, in a Nation that came into being with one of its principles being the widespread ownership of property? Ownership of property is one of America's cherished dreams, but this was dramatic evidence that few were achieving it if their dream involved farmland."[2]

These conditions deteriorated further after 1981. The Sherman Antitrust Act was seemingly ineffective. While riding my bus to the Pentagon on my way to the EPA, I met a woman lawyer who was expert on corporate monopoly. She worked in the Antitrust Division of the Department of Justice. She told me, in passing conversation, that monopoly in agribusiness harmed small family farmers. Yet she also admitted that breaking up large farms and agribusiness corporations was off the books. No one was able to stand in the way of large farmers and the industry. This was in the 1990s during the Clinton administration.

Outside the borders of America, the consolidation effect was just as bad as it was at home. My book, *Fear in the Countryside*, documented its detrimental consequences. Owners of plantations used tractors, fertilizers, pesticides, and science from America to speed up the removal of the small family farmers from their tiny pieces of land. *Fear in the Countryside* opened my eyes to injustice and corruption as well as corporate politics in America and the world. The book made my life difficult as I could never

---

[1] A. V. Krebs, *The Corporate Reapers: The Book of Agribusiness* (Washington, DC: Essential Books, 1992), 15–22.

[2] US Department of Agriculture, *A Time to Choose: Summary Report on the Structure of Agriculture* (Washington, DC, January 1981), 4, 6.

pretend to support corrupt science and bad policies. However, my words did not have the intended effect. I was working in bureaucracies that had their own rules and policies to follow. This led me to confrontations with senior officials – and trouble. Ironically, I ended up on Capitol Hill and with the EPA, which highlighted the profitable lives of industrialized agriculture, pesticides, and the taking over of rural America by plantation owners. *Fear in the Countryside* shaped my life and informed all my work. I spent about two years on Capitol Hill. I assisted in projects dealing with agriculture, food, and international development. Once again, I harvested the seeds I sowed in my book. I told my supervisors (in respectful and technical memoranda) that their bias against small family farming was misguided and against the national interest. In 1979, I joined the EPA. My responsibilities included environmental protection, especially in agriculture and the use of pesticides. I worked as an analyst for 25 years until 2004.

This new book, *Earth on Fire*, draws from my Greek and American experience. However, my theoretical vision of why agriculture should be small-scale and democratic originates in classical Greece. This was an era that gave birth to democracy, free and frank speech, the Parthenon, science, medicine, history, gorgeous art and architecture, craftsmanship, the Olympics, theater, poetry, literature, and the mind-boggling astronomical computer of genius, the Antikythera Mechanism. Agriculture was the motor and mother of poleis (city-states). Family farmers of each polis raised food. They were volunteer soldiers defending the freedom and independence of their small states. Agriculture was the polis, small family farms, and civilization. The period of decline started with the Roman conquest of Greece in 146 BCE. The Renaissance of the fifteenth century revived Greek learning but did not put a brake to the plantations worked by slaves. The industrial revolution of the nineteenth century consolidated the forms agriculture took in the twentieth and twenty-first centuries.

My work on Capitol Hill and with the EPA brought me face to face with the enormous scientific and political forces shaping rural America, including the variety and quality of food. I kept raising questions about the excessive power of the chemical and agribusiness industries and their lobbyists. I did not feel comfortable with the ambivalent responses of EPA scientists and senior officials to selfish corporate influences.

My historical access to classical Greece and the experience and knowledge I gained in my work merged. They illuminate this book. Some of the chapters are straight narratives, telling my story from a world

Smoke from the Gifford Fire fills the sky as the sun sets over Los Padres National Forest, California, August 4, 2025. (AP Photo/Noah Berger.)

I found disturbing. For example, pesticides are preliminaries and essentials to the transformation of family farms to animal factories. Honeybees have been suffering enormously from farmers' increasing use of pesticides. The harming of honeybees speaks of the tectonic shift in American attitudes and policies towards this extremely ancient and beneficial insect. The same attitudes threaten public and environmental health. Other chapters elaborate and go beyond arguments I made in earlier writings.

I illustrated the book with pictures that capture ideas and events relevant to the stories I told in the chapters of the book. One such image appeared in the World in Pictures by the Associated Press on August 4, 2025. The AP picture depicts a large red Sun over the red glow of a gigantic fire in the heart of California, the Gilford Wildfire. I personally found the message of this book, Earth on Fire, most strikingly portrayed by the photo "Smoke from the Gifford Fire fills the sky as the sun sets over Los Padres National Forest, California, August 4, 2025 by photographer Noah Berger featured by Associated Press" the World in Pictures. A quick Google search of AP the World in Pictures, Gilford Fire, August 4, 2025 will bring up the photo in question.

The 2020–2022 COVID-19 pandemic refocused my interests back to the human destruction of the natural world, potential mishaps at biological warfare laboratories, and, primarily, to factory animal farms. There are several thousand animal feeding operations throughout America, all sources for pathogens capable of exploding into pandemics. To understand these "farms" and their makers, I shed light on America's food and agriculture, revealing the forces that brought about the rapid decline of family farming and the control of rural America by corporate giants.

This is an age of conflicts and wars. Agriculture has joined the fray. It has been industrialized with huge machinery and deadly chemicals. Furthermore, animals by the millions are forced to live in extremely confined spaces that border on cruelty. These animals can become weapons of diseases – sometimes spreading diseases to one another and their caregivers.

But other times, like in the 2020–2022 COVID-19 crisis, animal disease could potentially become a pandemic. Americans, especially in early 2022, were still living in the great anxiety and hardship and death of the coronavirus pandemic. They want to hear the truth of why their world is falling apart. On May 5, 2022, the World Health Organization reported about 15 million deaths from the pandemic, of which some 930,000 were Americans.[3]

Pesticides are part of this outbreak of 2020–2022. They are at the heart of business agriculture. These chemicals alone: their use, misuse, and effects on nature, drinking water, and human health, constitute a great destructive force.[4] They encourage the extinction of wildlife and the slow but steady decline of human beings: fewer and fewer small family farmers, less democracy, food increasingly toxic, and more and more disease, including pandemics. Pesticides are the deleterious weapons that have forced honeybees and small family farmers out of rural America.[5]

---

[3] Benjamin Mueller and Stephanie Nolen, "Death Toll During Pandemic Far Exceeds Totals Reported by Countries, W.H.O., Says," *New York Times*, May 5, 2022.

[4] Robert van den Bosch, *The Pesticide Conspiracy*, with Preface by Paul Ehrlich (Berkeley: University of California Press, 1989).

[5] Carol Van Strum, *A Bitter Fog: Herbicides and Human Rights* (San Francisco: Sierra Club Books, 1981); E. G. Vallianatos with McKay Jenkins, *Poison Spring: The Secret History of Pollution and the EPA* (New York: Bloomsbury Press, 2014).

Journalism and information media rarely, if ever, connect farming with the arts of civilization. Yet agriculture, as I demonstrate in Chapter 1 about classical Greece, gave birth to civilization.

For a limited time, America had the great opportunity of maintaining the vital connections between land and civilization; however, today, only China is campaigning on ecological civilization. This is the result of another cultural revolution from the marriage of Chinese communism with American capitalism. So far, China is a world factory of material goods, which is slow-moving and chocking China with pollution; Chapter 2 explains the predicament of China in its struggles with references to ecological civilization.

Throughout this book I suggest ways farming can return to its original mission – including recapturing rural America from the oligarchs of banks, petrochemicals, fertilizers, machinery, meat, corn, and soybeans. I treat agriculture as the powerful social, political, and agrarian ways of sustaining people with food and culture. Agriculture was civilization – for millennia.

The 2020–2022 pandemic has had tremendous adverse effects, in a sense, it paralyzed and shut down dozens of countries, including the United States. Medical and government authorities offered no convincing explanation of the origins of the pandemic. This book directs the national and international conversation to a potential major source of the outbreak: conventional farming dependent on toxic sprays and animal farms. That in itself, I like to believe, would be a great contribution in helping America and other countries reform or abolish the agribusiness Cyclopes, which have caused so much harm for so many decades.

Evaggelos Vallianatos
July 2024

1

The Origins of Agriculture and Civilization: Classical Greece

Ikarios, a follower of the god Dionysos in a third century Mosaic from Paphos, Cyprus. According to lore, those who drank his wine were so intoxicated they killed him. The Greeks believed Zeus placed Ikarios in the constellation Bootes. Image from Public Domain.

Prologue

The temptations of the "modern" age are often irresistible. The seemingly pretentiousness of the missionaries and evangelists of modernity has no limits. Modern apologists cite the "modern" discoveries of science, glossing over the scientific discoveries of Aristotle and his colleagues. While the moderns admit, however reluctantly, that burning fossil fuels is warming the Earth, they are seemingly not yet willing to face the emergency of climate change by eliminating fossil fuels. In other words, the defenders of this age would rather continue with business as usual. They do not take the word "ecology" seriously. So, it's not surprising that the moderns praise the mechanical farm. After all, who wants to work hard and be at the mercy of insects and weeds?

Education

I am steeped in the literature, science, and philosophy of ancient Greece. The other reality is that I have lived in America for most of my life. My education at the University of Illinois, the University of Wisconsin–Madison, and Harvard University included science, history, and the history of science. I worked for America's political and environmental government for almost three decades. That experience shattered any illusions I might have had on the benefits of modernity. It taught me rather forcefully the limitations of government, science and industrialization.

Even my home village is Exhibit A of the global transformation that industrialization made possible. It breaks my heart when I revisit the Greek village that I was born in. These days, I see no donkeys, mules, and horses, the animals that were part of my early childhood and teenage years, and my father's agrarian world. Now small Japanese trucks and other small petroleum-fueled machines do the work of these vanished animals. While in the village, my sister Georgia – whose name in Greek means agriculture – and I go to the cemetery to light a candle in the grave of our parents. We then walk through the only olive grove still belonging to me. These trees of goddess Athena[1] no longer feel the same

---

[1]The connection between the olive tree and the goddess Athena comes from ancient Greek mythology. While exact first mentions are debated, the olive tree's connection to

as they did in my younger days. The bridge between my time spent in my homeland and my days in America from the 1960s to the present is simply too long.

Greece, especially the country and its civilization of ancient times, is my source for light and life. The source of life in the village and town is self-sufficiency in food. Aristotle, the fourth century BCE great pioneer in philosophy, political theory, and the inventor of science, said that a polis (city-state) has to provide its citizens with more than military security or a place for living. A state must afford its citizens ample opportunities to be self-sufficient and have a good life. Indeed, *autarkeia* or self-sufficiency in food is the purpose and happiness of a successful polis.[2]

The Shield of Achilles

Food self-sufficiency has had a very long history in Greece – all the way to the Bronze Age. In the eighteenth book of the *Iliad* of Homer, the god of metallurgy, Hephaistos, makes the Shield of Achilles on which he imprints images of heavens and Earth. The most prominent of the inhabitants of the sky are the tireless Sun god Helios, and the Moon waxing into her fullness. The constellations filling the heavens include the Pleiades (from sailing), the Hyades (from water), the mighty and spectacular Orion in the southern sky, and the Bear, which the Greeks also called the Wagon. The Bear, Homer says, turns about in place, looking back at Orion. The Bear alone remains aloof from the wash of the River Okeanos, the Ocean. This is because to an observer in the northern latitude, the Great Bear never goes below the horizon. Hephaistos added the great Ocean River to the outermost spherical rim of Achilles' Shield.[3] Homer's Bear or Wagon is Ursa Major. The Bear is a circumpolar constellation, turning around the celestial pole all night long without rising or setting. Homer knew that the Bear was also a guide to sailors. Odysseus, the god-like hero of Homer's

---

Athena became an enduring symbol in Greek culture during the Classical period (5th–4th centuries BCE).

[2]Aristotle, *The Politics* 1.2 (1252b27); 3.9 (1280b29–1281a2).

[3]Homer, *Iliad* 18. 486–489.

Odyssey, sails east by keeping the Bear to his left. Odysseus also sees the Dog Star (Sirius) of the constellation Bootes (from ox). Homer calls Sirius ἀστέρ ὀπωρινῶ, "*aster oporino,*" autumn star.[4]

At the center of the spherical Cosmos on the Shield of Achilles, Hephaistos depicted two poleis (city-states), one celebrating weddings and festivals, the other full of quarrels and fighting. But the most interesting and important pictures Hephaistos sculpted on his large and heavy Shield were images of rural life and self-sufficient farmers: the cultivation of rich farmland, farmers driving teams of oxen three times up and down the rows of the field. Each time the plowmen reached the end of the black land and turned, a worker ran up to them offering them a cup of sweet wine. Hephaistos portrayed reapers using sharp sickles harvesting grain from the field of the king. A feast of ox meat waited the harvesters and the king. Next to the grain harvest, Hephaistos painted a vineyard full of beautiful grapes. The grape clusters, supported by silver poles, were dark and the vines spread everywhere. Girls and boys harvested the grapes, putting the honey-sweet fruit in wicker baskets.

Next to the vineyard, Hephaistos gave life to straight-horn cattle. The cows were made of gold and tin. They run out of their barnyard to a field of grass near the banks of a wild river. Four golden herdsmen guided the cattle through swaying reeds. Nine nimble dogs followed the herdsmen. Two terrifying lions in the front of the herd had killed a bull and were eating its guts and black blood. The dogs in vain tried to frighten the lions. Hephaistos also made a pasture and a village full of silvery sheep, stables, roofed huts and stone animal pens. Hephaistos climaxed his rural delight with dancing young men and women. Young men were wearing tunics and golden knives hung from silver straps. The women were wearing delicate linens and were at the peak of their beauty. They danced with the young men by holding hands, hands clasping wrists.

Down at the Greek Farm

Homer is telling us the Greeks were primarily very small family farmers, intimately connected to the land and the growing of food. Agriculture for the Greeks was food, economics, natural history, science, religion and politics. In other words, agriculture was civilization. Homer reports that

---

[4]Homer, *Odyssey* 5. 272–277.

when Odysseus came in contact with the Cyclops Polyphemos, he faced an uncivilized barbarian. The Cyclopes, Homer writes, were: "Lawless savages who never plant crops or plow the land. Everything grows for them unsown, wheat, barley, and vines with clusters of grapes. Rain from Zeus waters these crops. However, the Cyclopes have neither assemblies for law making nor any established customs."[5] Indeed, according to Euripides, a fifth century BCE dramatic poet, Polyphemos was also full of hubris. The one-eyed monster says to Odysseus that he cared less about the gods. As far as he was concerned, the greatest god was his own stomach. The Cyclops bragged that eating and drinking for a day and causing no pain to himself was Zeus for men of good sense. Polyphemos had contempt for those who established laws, which complicate people's lives.[6]

Hesiod, the epic poet only second to Homer, focused his masterful poem, *Works and Days*, on farming, urging the Greeks to work the land for food, security, community, and for the favor and pleasure of the gods. The peasant had to know how to cultivate the land and take care of his trees, animals and household. He had to feel comfortable with growing of food and be competent to plow, hoe, sow, prune, harvest, and thresh his grains. He should also be able to construct his own plow and wagon and other farm implements, including choosing oxen for the cultivation of the land. In addition, Hesiod preaches the peasants should connect farming to natural phenomena and the Cosmos: working the land in accord with the seasons, the weather, the return of the migratory birds, the position and movement of the stars and, especially, the phases of the Moon. In addition, Hesiod envisioned the reign of justice in agriculture. "Men whose justice is straight and honest," he says, "know neither hunger nor ruin but, in the midst of feasts, they enjoy the yield of their labors. The Earth brings them a rich harvest: their oak trees teem with acorn and the middle with honeybees. Their fleecy sheep are weighted down with wool. Start reaping when [the constellation] Pleiades rise and begin to plow when they set."[7] The Greek peasant-farmer had to try being virtuous. He had to know enough about the heavens in order to be successful in raising his food. Water was intimately connected to that understanding of the heavens – and water has always been at the heart of agriculture.

---

[5] Homer, *Odyssey* 9.106–112.
[6] Euripides, *Cyclops* 332–340.
[7] Hesiod, *Works and Days*, 230–234; 383–384.

Sacred Water and Greek Cosmology

Water was sacred to the Greeks because nature and the Cosmos were sacred. The gods, and the universe those gods represented, demanded that their pious followers understand their power.[8] This meant primarily understanding nature, and the cause and effect of phenomena in the natural world and the universe. Mythology informed them that:

(1) The god of southern wind, Notos, was the father of rain; Zeus having given him the prerogative of sending rain-giving clouds from the sky to Earth.

(2) Hephaistos, son of Hera, was fire – the all-devouring, all-taming, and all-haunting part of the universe. Hephaistos – the god whose craftsmanship sparked metallurgy in the Aegean Island of Lemnos and the Asian mountain of Caucasus – was the eternal artisan who brings light to mortals who see the ether, the Sun god Helios, the stars, the Moon and pure light through him. Hephaistos in fact lives in mortals and, because of that, nature itself burns in their bodies. In other words, Hephaistos, the god of fire, was reason in the Cosmos and in the world of humans. He surpassed all the other gods in craftsmanship.

(3) Nomos, a cosmic principle, was all about reason and order in the Cosmos. It sets limits for life on Earth as much as it arranges the stars in the universe.

(4) Okeanos, the mighty river around the Earth, was where all life, including the life of the gods, came from. Okeanos was also the father of all seas, rivers, and streams of the Earth from his marriage to his sister Tethys.

Aristotle confirms that the ancient Greeks considered Okeanos and Tethys the parents of creation. The gods themselves took their oath by the water of the Styx, a river in northeastern Arcadia in Peloponnesos. Hesiod says that the gods would pay a terrible price if they swore a false oath by the primeval and immortal water of Styx. Water was at the heart of the Greeks' Cosmos. It became the key, which Thales associated with the origins of Cosmos.

---

[8] Hesiod, *Theogony*, 233–370; *The Homeric Hymns [to the Gods]*, tr. Apostolos N. Athanassakis (Baltimore: The Johns Hopkins University Press, 1976).

Thales, c. 624–545 BCE, was a Greek cosmologist from Miletos, a famous Greek Ionian polis on the coast of Asia Minor near the mouth of the river Maeander. Aristotle considered Thales the founder of natural philosophy. Thales made water the foundation of the universe; He said the Earth rests on water.[9]

Herodotos, the father of history who wrote the history of the wars between the Persians and Greeks, shares the prevailing idea that agriculture was a mark of civilization. He said that in Scythia (present day Ukraine and southern Russia) there was this community, the Kallippidai, in the middle of the coastline of Scythia. He also claimed that these Kallippidai were Greek Scythians who in all manner of customs were like the rest of Scythians, save for their food. The Greek Scythians "cultivate the land and ate grain, onions, garlic, lentils, and millet." Herodotos said there were other Scythians that cultivated grain only to export it while others made a living by farming. There were also nomadic Scythians who "do not grow food or cultivate the land."[10] In the eyes of Herodotos, those Scythians who, like the Greeks, cultivated the land and ate grains, were preferable to the nomads. They were civilized.

Finally, Xenophon, c. 428–354 BCE, a student of Socrates, a general, and historian, much like Herodotos, loved farming. He believed that agriculture was the foundation of Greek society. Peasant farmers were, by necessity, the best of citizens because they had the skills and courage to defend the freedom of the polis. He was certain agriculture was the mother and nurse of the arts of civilization.[11]

The Gods and the Divine in Agriculture

Greek agrarian civilization combined polytheism, theater, athletics, beautiful temples, democracy and science. But the gods were at its very foundations. Plato was certain gods existed, though he suspected we know practically nothing with certainty about them. According to him, the gods guarded virtuous citizens and were beyond corruption.[12] The tragic poets,

---

[9] Aristotle, *On the Heavens* 294a28–33.

[10] Herodotos, *The Histories* 4.17–19.

[11] Xenophon, *Oeconomicus* 5.7–14; 17, 20, ed. and tr. by Sarah B. Pomeroy (New York: Oxford University Press, 2001).

[12] Plato, *The Laws* 10.907.

Homer and Hesiod, praise the gods. Euripides captures that great consensus in saying "Greeks don't trifle with divinity." Their fathers passed down to them ancient and sacred customs and traditions that no modern logic could overthrow. Wine, for example, was as much a gift of the god Dionysos to the Greeks as grains were the gift of goddess Demeter. When we pour wine in our libations, Euripides says, we pour Dionysos himself, so through his intercession we may win the favor of the gods.[13] To this day, Greeks call grains like wheat and barley *Demetriaka*; gifts of Demeter.

The Greeks fished and traded and manufactured things. But they depended on the land. They had to know the best time to plant and the best time to harvest their food. They had a star calendar, which listed the risings and settings of the stars in relation to the Sun god Helios. The constellation Pleiades, for example, emerged just before dawn in the east in late May; it was as good an astronomical message of impending wheat harvest as one could get. Hesiod had stressed the reliability of that phenomenon. This sighting also meant the beginning of summer. When Arcturus, the brightest star in the constellation Bootes, rose, the Greeks knew that autumn was with them. Like the Olympian gods, these stars were gods to the Greeks. So, the gods and the Greeks were inseparable, especially in the working of the land and everything to do with the formation of the political institutions of Greece. As classical scholar Mary Lefkowitz rightly observed, the main advantage of Greek religion is that "it describes mortal life as it really is: fragile, threatened, uncertain, and never consistently happy... It is a religion for adults, and it offers responsibilities rather than rewards."[14]

Farming and Public Virtue

One of those responsibilities was to be virtuous in their personal and political lives. Farming was the key to public virtue. Plato was no peasant, but he understood farming kept society together. He said the wise legislator must not allow any citizen to own a farm which was more than four times larger than the average-sized farm, which was about five acres or less.[15] The size of the farm was not a technicality. In early sixth

---

[13] Euripides, *The Bacchae* 200–285.

[14] Mary Lefkowitz, *Greek Gods, Human Lives* (New Haven: Yale University Press, 2003) 239.

[15] Plato, *The Laws* 5.744.

century BCE, Athens faced an existential crisis because the size of the farms had been left unchecked: large farmers enslaved the peasants. Athenian nobles, fearful for their survival, invited a former archon (top leader), Solon, to reorganize the Constitution. Solon abolished slavery and forgave all farm debts. He saved Athens from class warfare and tyranny.[16] Solon's reforms eventually led to Athenian democracy. Many small farms balanced the power of large farmers and preserved internal peace and, therefore, the practice of the worship of the gods. To the Athenians, these divine beings were not merely in magnificent temples. They were primarily in the crops and fields of the Greeks.

Zeus, father of the Greek gods, controlled rain and thunder. Athena, born out of the head of her father Zeus, gave the Athenians the olive tree. This tree was so important to the Athenians that Athens prohibited its destruction.[17] Another son of Zeus, Dionysos, brought to Greece the grape vine and wine. Ikarios was the first Greek who made wine, though his enthusiasm for the sweet drink cost him his life.[18]

Pan cared for the flocks of sheep, goats and cattle, and Artemis protected the natural world. Aristaeus, son of Apollo, protected honeybees, shepherds, olive trees, and the making of cheese.

Demeter: Queen of the Threshing Floor

Demeter, sister of Zeus, was the goddess of wheat. The *Homeric Hymns*, a collection of ancient Greek poetry dedicated to the gods, say Demeter was "the bringer of the seasons," and "the greatest benefit and joy to undying gods and mortals."[19] The third century BCE Greek poet Theocritus of Syracuse thought of Demeter as "the Queen of the threshing-floor," whose hands were "laden with poppies and sheaves" of wheat.[20] She and her daughter, Persephone, goddess of spring, sent a young prince, Triptolemos, around the world teaching people the art of agriculture. The Greeks gave

---

[16]G. R. Stanton, ed., *Athenian Politics c. 800–500 BC: A Sourcebook* (New York: Routledge, 1990) 34–85.

[17]Ilias Arnaoutoglou, ed., *Ancient Greek Laws: A Sourcebook* (New York: Routledge, 1998) 35–36.

[18]Apollodorus, *Bibliotheca* 3.14.7; Nonnus, Dionysiaca 47.34; Hyginus, De Astronomica 2.4.4.

[19]*The Homeric Hymns*, 54. 268–269.

[20]Theocritus, *Idylls* 7.6–8, tr. Anthony Verity, intr. Richard Hunter (Oxford: Oxford University Press, 2008).

Fig. 1. Silver coin of 4 drachmas from the Cycladic island of Syros dating from the mid-second century BCE with the image of Demeter on one side. Notice the sheaves of wheat and small snakes included in her hair. Goddess Demeter was Earth herself. Courtesy of the Numismatic Collection, Alpha Bank, Athens.

Demeter special honors. They celebrated her in the Eleusinian mysteries in Eleusis near Athens. This was their greatest religious festival. It took place during the sowing of the crops so that the honored gods, Demeter and Dionysus, would bless those crops (Fig. 1).

This great Panhellenic tradition lasted for millennia. But the Christianization of the Roman Empire in the fourth century was like a massive Earthquake that wrecked Greece, including the Eleusinian mysteries.[21] Yet not everything was lost. Fragments of the ancient Greek traditions remained in the minds and character of the Greeks for centuries. For example, a statue of Demeter in Eleusis survived the barbarian and Christian holocausts and Turkish occupation until 1801. Once the Greeks lost their freedom and the country became a province of Rome in 146 BCE, northern European Germanic barbarians, the Heruli, in 267 of

---

[21] Evaggelos Vallianatos, Christians and the classics: War against reason, *Mediterranean Quarterly,* Summer 2004, pp. 75–94.

our era, sacked Athens. However, the forced Christianization of Hellas nearly remade Greek culture. This upheaval started in the fourth century of our era and lasted for several centuries. The Christian holocaust was more lasting than the barbarian invasions and plunder. Thus, in 1801, the farmers of Eleusis were so convinced Demeter ensured the prosperity of their crops that they protected the statue with their lives. In 1801, however, British officials looted the statue of Demeter. It is now in a museum in London.

Greek farmers struggled for centuries in keeping alive their contact with their ancient ancestors. In modern times, the Greeks love attending the performance of ancient tragic plays in the surviving Greek theaters. Those tragedies speak to them intimately.[22]

The methods of farming that Homer describes were pretty much the methods of Greek farming down to the twentieth century. My father worked within the Homeric farming tradition. The supreme role of bread and wine in Greek diet, the near sacred state of hospitality, and the intimate role of the divine in farming (the gods in ancient times and, to some degree, Christian saints in the medieval and modern eras) constitute an almost unbroken continuity in Greek civilization.

What to Do

In the lives of civilization and agriculture in the twenty-first century, not much of antiquity has survived unchanged. The Greeks thought of animals as nearly sacred – certainly indispensable for human existence and culture. But now animal factories, like the remaking of crops by genetic engineering, are no longer sacred. Ancient Greek farming is a mine of ecological and political insights for thinking through an alternative way of raising food without deforming the world and civilization. Combine the Greek passion for small and democratic family farms with restored ecosystems, the sacredness of the land and the natural world, and you have a sustainable new ecological agriculture capable of fighting global warming, pandemics, and feeding the world.[23]

The next chapter examines agriculture and civilization in China, which like ancient Greece, depended on farming for living and culture.

---

[22] George Thomson, The continuity of Hellenism, *Greece and Rome*, April 1971.

[23] E. Vallianatos, The democratic and sacred nature of agriculture, *Environment, Development and Sustainability* (2012), 14: 335–346,

2

China: Iowa Farming or Ecological Civilization?

A Chinese peasant selecting seeds.
Courtesy: Beijing Center for Ecological Civilization.

Deep Passion for the Land

In late October to early November 2014, I visited China for the first time to attend a conference in Beijing. That conference was no typical academic conclave full of abstractions and jargon. Instead, it was a continuing Chinese discussion on the values of ecological civilization. Yes, indeed, ecological civilization! These eco-political discussions have been taking place in a city, Beijing, and a country, China, which are the workshops of the world. There's nothing ecological about that.

During my first three days in Beijing, I could not see the sky. A white-brown haze engulfed the skyscrapers and Beijing. Otherworldliness was in the air. The streets were full of cars and taxis. I stayed in two hotels; one was attractive. Each hotel building was modest in size and had the feeling of community. In the first hotel, I walked a long corridor every morning for breakfast. The corridor itself was an exhibit of fashionable clothing and art. The cafeteria was full of foreigners. The other hotel was a skyscraper. From my 12th story room, I watched modern Beijing: a gigantic city of concrete built on the models of New York and Chicago.

Eventually, I saw the sun and Beijing became friendlier. I noticed trees on both sides of highways and city streets. I even found a fruit stand very close to my hotel. But my greatest discovery was seeing buildings and museums preserving ancient Chinese culture and architecture. Once in such rare surroundings, like the Guo Moruo Memorial Museum and the former residence of Soong Ching Ling[1] (turned museum), I felt like I was in the midst of Greek antiquities. The beautiful buildings and their gardens gave me aesthetic pleasure and brought me close to Chinese history and the natural world.

A Chinese professor of philosophy and I used the convenient and clean Beijing subway to visit the fabulous Forbidden City: the governing palaces of the twenty-four emperors of the Ming and Qing dynasties lasting from the early fifteenth to the early twentieth centuries. Thousands of Chinese filled the large courtyards between the royal palaces. I had the sensation that the Forbidden City was more than a museum. It was a place of history and respect for ancient Chinese civilization, and a temple

---

[1] Soong Ching Ling, a Chinese political figure and the third wife of Sun Yat-sen (then Premier of the Kuomintang and President of the Republic of China) was often referred to as "Madame Sun Yatsen."

of worship. My friend and I passed through the modern center of Chinese government, the immense Tiananmen Square or Gate of Heavenly Peace. We then sampled the incomparable Chinese food: a feast of fish, rice, and vegetables.

Rural China

I had an opportunity to visit two villages on the periphery of Beijing. A professor of rural sociology from the China Agricultural University drove a couple of Chinese colleagues and me to his brother's six-acre persimmon farm. The fruit, persimmons, red in color and soft as large heirloom tomatoes, were out of this world. I ate a couple of them and tasted ambrosia, the food of the gods. The small farm owner and his father were self-sufficient in food. They said they had a good life on the land. They sold their fruits to Beijing grocery stores and people from the city. Their affection for their farm was strong, as they inherited it from their ancestors. Once I said I was in China for a conference on ecological civilization, they smiled and said the Chinese had deep passion for the land.[2]

The other village I visited was a horticultural paradise. Its trademark is the humble cabbage. A large statue of the cabbage surrounded by watermelon slices and statues of farm animals adorned the entrance to the village. I saw a series of greenhouses growing tomatoes, mushrooms, and other vegetables. This is a village where the land is held in common. The president of the village explained how each family does work in a spirit of equity and justice. He spoke eloquently on farming and defended rural civilization. He then treated us to great food. It's this village reality – the Chinese farmers' deep passion for land – that makes sense of any discussion in China about constructing or reinventing an ecological civilization.

Yet reality is often deceptive, hiding historical forces that shape the present. Like the United States, China embarked on its modernization by wrecking rural society. The Marxists were not that much better than the capitalists. But, fortunately, China started much later than the United States in its path of modernity and industrialization. This means about

---

[2] Evaggelos Vallianatos, "Deep Passion for the Land – in China," *Truthout*, November 13, 2014.

260 million Chinese are still small farmers working the land. While China has embraced industrialized agriculture, there is still room for the protection of rural society. That is one of the reasons that the Chinese are debating the advantages of ecological-rural civilization. I heard several farmers explain why they do what they do. I heard stories of saving villages from extinction. The difficulties, however, persist. Young people keep moving to the cities. Farmlands become real estate. Entire villages become cities. And the president of China, Xi Jinping, is seemingly making limited progress in reducing poverty in the country: farmers from inaccessible villages are relocated to new settlements with water, electricity and new roads while leasing out their lands to hotel and resort developers. However, some of the villagers did not find improvement in their new lives and still struggle with finding work to earn a living.[3]

I defended China's traditional farming and ecological civilization at several universities in Beijing. I spoke at the China Agricultural University, Minzu University of China, Beijing Normal University and the China University of Political Science and Law. In each case, I praised the talent and perseverance of China's rural people in raising food and protecting the natural world. My modest proposal was to introduce agroecological science to further strengthen the food self-sufficiency and food security of China. This is good for China and the entire world.

Ecological science confirms the science and wisdom of small-scale farming. This means there is no need to convert farmlands into the practices of industrialized agriculture. Small-scale farmers compete nicely with large, industrialized farmers. In fact, small family farmers are much more productive than large farmers.[4] They employ and benefit from traditional knowledge and local talent. Their practices of growing a variety of local crops nurtures biological and genetic diversity, thus nurturing sustainability and food sovereignty. Moreover, smallholders are gentle on the land, enriching its soil and fighting climate change. The food they raise is wholesome and nutritious. This makes ecological civilization possible.[5]

[3] Alice Su, "In China, Ending Poverty Costs the Poor," *Los Angeles Times*, November 27, 2020.

[4] Miguel A. Altieri, "Agroecology, Small Farms and Food Sovereignty," *Monthly Review*, July–August 2009.

[5] Evaggelos Vallianatos, "Ecological Civilization: Could China Become a Model for Saving the Earth?" *Counterpunch*, March 27, 2019.

Such a prospect could open the path not taken: merging traditional agrarian knowledge with ecology to feed ourselves and prevent the additional deterioration of ecosystems. China has started this important discussion. We must do the same. Deep passion for the land is a universal virtue.

Modernizing Rural China?

The government of China is seemingly attracted to the concept and reality of becoming like the United States. Modernity, however, appears to be causing a rift in Chinese society. Left unchecked, modernity could trigger the dissolution of the surviving traditional Chinese culture and potentially, even breaking up the state.

Modernity is threatening sustainable agriculture in China. It represents forces that could bring traditional agrarian civilization to an end. These forces include industrialized farming tied to large farms, businesses, large machines, the separation of animals from crops, the growing of only one crop per field without rotations, and the use of vast quantities of herbicides, synthetic fertilizers, and biocides.

The opening of China to America in the late 1970s led to the influx of Chinese studying agriculture at American agricultural or land grant universities. This accelerated the modernizing policies of the Chinese government. But if American agribusiness ideology and practices would take hold of the Chinese government's modernization policies, what would happen to the millions upon millions of small family Chinese farmers? While family farmers are not perfect, they operate within the wisdom and restraint of ancient traditions. Nature is still their book.

Yet ancient China left evidence of bad agricultural practices with dreadful ecological consequences. For example, the people who lived at the Loess Plateau "devolved" to the point they destroyed its land. This is no small farm but a vast region of 640,000 square kilometers in northwestern China – approximately the size of France. The degradation of the land lasted for more than a thousand years. Annual floods dumped 1.5 billion tons of soil into the Yellow River. In 1995, the Chinese government borrowed US$500 million from the World Bank and rehabilitated a section of the Loess Plateau about 35,000 square kilometers in size. In ten years, the re-vegetation of the land brought back functioning ecosystems comprising biodiversity, circulating water, soil stability, natural fertility and a healthy environment sequestering carbon from the atmosphere.

Authorities forbade logging, the growing of crops on steep slopes, and the grazing of goats and sheep. About 2.5 million Chinese peasants benefited from a healthier Loess Plateau.

The story of the Loess Plateau comes from John D. Liu, Chinese American filmmaker and ecologist. He asks that we rethink our traditions, especially those of land tenure and economic development. Holding on to ways invented long ago but untested by logic, science, and justice is harmful, he says. Biodiversity loss and climate change, according to Liu, demand that, "humanity evolves to a higher level of consciousness or knowingly chooses to cause future generations great harm." Unless we find the courage to mend our relationships with the Earth, putting a steep price on pollution and environmental destruction, and giving proper value to fertile land, wildlife, biodiversity, rivers, uncontaminated water, clean air, forests, and a healthy environment – what he calls "ecological function" – we are headed for intolerable trouble and, possibly, extinction. Liu documented the decade-long ecological recovery of the Loess Plateau. He is convinced it was degraded by exploitation. Chinese medicine and philosophy, he says, preached the blessings of the natural world. However, he claims that "the daily reality in China for thousands of years has been much less respectful to nature." While nature thrived in the Loess Plateau, Liu says, it was a cradle of Chinese civilization. This was the home of the Han Chinese and the power center of the Han, Qin, and Tang dynasties.[6]

Liu is convinced that "unsustainable agricultural practices" are at the root of ecological degradation. He urges us to abolish such practices.[7] In fact, Liu equates modern industrialized farming to Neolithic agriculture with tractors and toxins.[8] Liu is very excited over the ecological healing of a large segment of the Loess Plateau. He has become a messenger to the world that time has come for the powerbrokers to mend their horrible ways of destroying the Earth. Lessons he learned in the recovery of the Loess Plateau could be applied everywhere in the healing of degraded land, and much more. "The fundamental lesson of the Loess Plateau rehabilitation," he says, "is that it is possible to rehabilitate large-scale

---

[6] John D. Liu, "A Continuing Inquiry into Ecosystem Restoration: Examples from China's Loess Plateau and Locations Worldwide and their Emerging Implications," in *Land Restoration: Reclaiming Landscapes for a Sustainable Future* (London: Elsevier, 2015), pp. 361–379.

[7] John D. Liu, "Earth's Hope: Learning How to Communicate the Lessons of the Loess Plateau to Heal the Earth" (Environmental Education Media Project, 2007).

[8] John D. Liu, "Is it Possible to Rehabilitate Large-scale Damaged Ecosystems" (Permaculture Research Institute, June 29, 2016).

damaged ecosystems – including those that have been degraded over the course of centuries or even millennia. This is of enormous importance given the huge areas of the Earth that have been degraded by humans since the advent of settled agriculture, and the emerging risk from human-induced climate change.... What if these areas could be restored? What would restoring the Earth to productivity over vast areas mean in terms of mitigation and adaptation to climate change, availability of food, economic security, social cohesion, and even military security?"[9]

I agree with Liu and share in the wisdom and excitement of his questions.

Imagine a world with no more pesticides and factory farms, and no more taking of wildlife in the oceans and land? What would that mean to today's government-sanctioned plunder of the planet? Liu is relatively optimistic about China. In a message on November 21, 2017, he said to me: "The Chinese have a two-track development plan ... they are developing their economy to improve living standards for the people, and at the same time they are working to create a harmonious ecological civilization. My conclusion is that biodiversity, biomass and accumulated organic matter are the indicators and determinants of ecological function. Somebody very high up in the Chinese government has realized this and it is enshrined in law and action. Over the nearly 40 years that I've been closely observing China's development, this has been a constant trend."

Liu may be right. However, agricultural modernization in China is backfiring.

Pesticides are decimating honeybees worldwide. Yet demand for honey is soaring. China has been taking advantage of this tragedy by exporting huge amounts of cheap but "tainted" honey.[10]

China also seems to be relying on genetically modified (GM) soybeans from America for cooking oil. These GM soybeans are drenched with Monsanto's controversial weed killer glyphosate, a probable carcinogen and a straightforward biocide. Chen I-wan, advisor to the Committee of Disaster History, China Disaster Prevention Association, is

---

[9] John D. Liu, "A Continuing Inquiry into Ecosystem Restoration: Examples from China's Loess Plateau and Locations Worldwide and their Emerging Implications," in Land Restoration: Reclaiming Landscapes for a Sustainable Future (London: Elsevier, 2015), pp. 361–379.

[10] "Asian Honey, Banned in Europe, is Flooding U.S. Grocery Shelves, *Food Safety News*, August 15, 2011. See "Rotten," a Netflix documentary: "Lawyers, Guns and Honey," January 2018.

very unsettled by China's dependence on Monsanto. He says Chinese medical scientists, the Chinese Soybean Industry, and senior military officers are convinced that the Monsanto soybean oil is giving Chinese citizens diseases – perhaps even cancer.[11]

Ye Jingzhong, professor at the China Agricultural University in Beijing, is troubled by the government's policies of pushing industrialized agriculture throughout the country. He asks, "Is it possible to imagine China without traditional farmers?" He cited the 2011–2015 National Modern Agricultural Development Plan as having unambiguously modern ambitions. He wrote, citing the Plan: "China needs to equip agriculture with modern material conditions, renovate agriculture with modern science and technologies, increase agricultural production with modern industrial systems, enhance agricultural production with modern business methods and guide its development with modern development concepts and newly born farmers."[12]

However, the agrarian reality in China clashes with the development strategy of the government. The government's policy favoring the "modern" is preposterous. It almost nullifies its calls for an ecological civilization. Is it preparing the ground for a potential confrontation with the country's peasant-farmers? According to Ye Jingzhong,[13] there are 260 million peasant-farmers growing sufficient food to feed China. In fact, he reports that food production by the peasant-farmers has been increasing for several years. Yet small farmers often have difficulty in selling their grains. So, why has the government been promoting industrialized farming? What are the hundreds of millions of peasant-farmers supposed to do? The Chinese government's failure to answer these questions clouds the future of the entire country. Yet the government's intention is clear. It is funding modern factory agriculture. Some of the world's largest industrialized farms are in China. This is especially true for meat and dairy. The Chinese government owns and operates most of the gigantic high-tech farms.[14]

The Chinese government is also spending billions on scientific research and experimentation in imitating how America grows its crops.

---

[11] Chen I-wan, personal communication, May 19, 2014.

[12] Ye Jingzhong, "Land Transfer and the Pursuit of Agricultural Modernization in China," *Journal of Agrarian Change*, Vol. 15, No. 3, July 2015, pp. 314–337.

[13] *Ibid.*

[14] Tracie McMillan, "Feeding China," *National Geographic*, February 2018, pp. 88–107.

Fig. 1. State-owned animal farm in China, identical to the animal farms of Iowa, US. *Courtesy*: Chen I-wan.

Chinese scientists claim they are boosting yields but limiting the damage associated with modern farm practices. Yet, no matter the sophisticated process of moving results from plots to farms, testing for higher yields, "crop uptake of nutrients," and genetic engineering of crops, the effects are bad for the environment and for people eating such food.

Wu Kongming, vice president of the Chinese Academy of Agricultural Sciences, says Chinese agriculture is suffering from "excessive use of chemical pesticides and fertilizers."[15] Zhu Youyong, professor at the Yunnan Agricultural University, complains that, during the last century, modernization has been responsible for a dramatic decline in crop varieties in China. For example, rice varieties went from 46,000 to 1,000. He is convinced cultivating just one crop and losing "farmland biodiversity" result in "pest and disease epidemics" that, in their turn, cause "overuse

[15] Wu Kongming in *Report of the International Symposium on Agroecology*.

of pesticides" and increase the risks from eating such food. Professor Zhu Youyong also says that crop diversity reduces disease and increases yield.[16]

Chinese and foreign scientists are transforming Chinese traditional farms into Iowa-like fields. They would like to see their work become a model for the world.[17] It's unlikely, however, that the world is interested in how China is imitating American farming practices. No matter the quality of minute science in calibrating the crops for higher yields, that is not how the crops flourish in nature.

Diversity of crops in rotational basis helps to fight diseases and also produces better harvest. Biodiversity is the key. Reducing or losing agricultural biological diversity, however, is catastrophic – it is leading us to a "silent, rapid, inexorable" rendezvous with extinction; to the doorstep of hunger on a scale we refuse to imagine. To simplify the environment as we have done with industrialized agriculture is to destroy the complex interrelationships that hold the natural world together. Reducing the diversity of life, we narrow our options for the future and render our own survival more precarious. It is life at the end of the limb.[18]

Modern farmers and scientists, however, ignore biodiversity. They prefer growing one crop at a time, fighting the corresponding diseases with deleterious pesticides. The world cannot afford monoculture and the blind science that comes with it. The world cannot afford to go on producing food without its farmers. It will probably starve without them.[19] The Chinese government is not completely ignoring the ecological and social pathologies of modern farming. It is trying to relink China to its ancient ecological culture by sponsoring a national and international conversation

---

[16] Zhu Youyong in *Report of the International Symposium on Agroecology.*

[17] Fusuo Zhang, Xinping Chen and Peter Vitousek, "An Experiment for the World," *Nature*, **497**, 2 May 2013, 33–35.

[18] Cary Fowler and Pat Mooney, *Shattering: Food, Politics, and the Loss of Genetic Diversity* (Tucson: The University of Arizona Press, 1990), p. ix.

[19] Evaggelos Vallianatos, *Fear in the Countryside: The Control of Agricultural Resources in the Poor Countries by Non-Peasant Elites* (Cambridge: Ballinger Publishing Company, 1976); *Harvest of Devastation: The Industrialization of Agriculture and its Human and Environmental Consequences* (New York: The Apex Press and Goa, India, 1994); *This Land is Their Land: How Corporate Farms Threaten the World* (Monroe, Maine: Common Courage Press, 2006).

on ecological civilization. The strategy may be to convince the Chinese to pollute less.

Pollution and Traditional Knowledge

Pollution in China is life threatening. In 2005, Pan Yue, deputy director of the State Environmental Protection Administration of China, warned that the so-called Chinese "miracle" of being the workshop of the world was not much of a miracle. He said: "[The Chinese miracle] will end soon because the environment can no longer keep pace. Five of the ten most polluted cities worldwide are in China; acid rain is falling on one third of our territory; half of the water in China's seven largest rivers is completely useless; a quarter of our citizens lack access to clean drinking water; a third of the urban population is breathing polluted air; less than a fifth of the rubbish in cities is treated and processed in an environmentally sustainable manner." In addition, Pan Yue connected ecological degradation to the economy: "[The effects of abusing the natural world are] massive. Because air and water are polluted, we are losing from 8–15% of our gross domestic product. This does not include the costs for health and human suffering: in Beijing alone, 70–80% of all deadly cancer cases are related to the environment. Lung cancer has emerged as the number one cause of death."[20]

Cancer and other human debilitating and deadly diseases are intimately related to the state of the environment.[21] Human and environmental health are one. Damage one and you will find the roots in the other. China is the mirror of human hubris, the unthinking and rapacious idea of the control of the natural world. China is living the effects of a "severe ecological crisis."[22] This reality and ecological devastation associated with modernization may have triggered the pedagogic effort of the Chinese government in spreading the idea of ecological civilization among the Chinese people. James Oswald, an Australian researcher working for the Chinese government in Beijing, is convinced the campaign of

---

[20] Andreas Lorenz, "China's Environmental Suicide," *Open Democracy*, April 5, 2005.

[21] Janet D. Sherman, *Life's Delicate Balance: Causes and Prevention of Breast Cancer* (New York: Taylor and Francis, 2000); Samuel S. Epstein, *Cancer-Gate: How to Win the Losing Cancer War* (Amityville, New York: Baywood Publishing Company, 2005).

[22] Zhihe Wang, Huili He, and Meijun Fan, "The Ecological Civilization Debate in China," *Monthly Review*, **66** (6), November 2014.

the government on ecological civilization is designed to change individual consciousness and behavior for some kind of transition to a potential "sustainable environmental future." This message is probably sinking in. On October 12, 2017, at Beijing Normal University, I listened to Xu Jialu, vice president of China, speak with knowledge and passion about ecological civilization and why China must change and defend the natural world from more pollution and why traditional wisdom matters. Xu is probably the most senior government official who tempers the tenacity of the modernizers in China.

Another encouraging political and ecological development is the rise of ecological farmers in Chinese cities. I had the opportunity to visit an unusual farm at the outskirts of Beijing. This was one of many farms trying to feed the residents of Beijing good food. This kind of farm, growing fruits and vegetables without pesticides and genetic engineering, was familiar to me because it was like the farm in California from which I receive a box full of organic fruits and vegetables every week. The same thing is happening in Beijing. The founder of Share Harvest cooperative farm, Shi Yan, has a doctorate in ecology. She said she started small. Several Chinese professors funded her experiment. Now her farm feeds around 2,000 families in Beijing. The farm is about two hours by car from Beijing.

In the United States we say these cooperative farms make up "community-supported agriculture." They do the same thing in China. However, neither this hopeful Chinese farm movement, nor organic traditional small-scale farming suffice to put the brakes on the "modernity" complex of China.[23] Xu Jialu's influence is probably extensive, but China needs much more official and public environmental protection. Environmental degradation in China is of such colossal dimensions that it threatens the survival of the State. This may be the reason why the Chinese government embarked on a systematic campaign on behalf of ecological civilization. According to Oswald, the Chinese government is pushing to convince the Chinese people that ecological civilization is good for them, possibly converting them into "ecologically civilized environmental subjects."[24]

---

[23] Evaggelos Vallianatos, "Growing Ecological Civilization in China," *Counterpunch*, November 8, 2019.

[24] James Oswald, "Environmental Governance in China: Creating Ecologically Civilized Subjects" (Ph.D. thesis, University of Adelaide, 2017), pp. 1–7.

Fig. 2. Community-supported agriculture: share harvest farm workers boxing vegetables for delivery to Beijing residents. Photo by Evaggelos Vallianatos.

Another Australian scholar, Bill Laurance, professor at James Cook University, Australia, is charging China with global ecocide. He says China is investing so much in the Belt and Road infrastructure project all over the world that its ecological footprint is threatening to trample the natural world.[25] This new "silk road" connects China to Europe, the rest of Asia, Africa, and Latin America. By 2019, China had invested about US$440 billion in opening roads and sea lanes for her goods to reach the world. Some critics see all this "development" as "debt traps" for poor countries and another sophisticated method for rising Chinese influence and geopolitical reach in world affairs.[26]

[25]Bill Lauramce, "China's Growing Footprint on the Globe Threatens to Trample the Natural World," *The Conversation*, December 5, 2017.

[26]Alice Su, "In China, Rebranding a Foreign Loan Initiative to Calm Fears Over Geopolitical Reach," *Los Angeles Times*, April 26, 2019.

A Chinese critic, Yi-Zheng Lian, commentator on Hong Kong and Asian affairs, says China is making itself a superpower. He speaks of Chinese communist leaders from Mao Zedong to Xi Jinping having a Chinese dream to conquer the world. Mao won the confidence of farmers in the countryside. Then he surrounded the cities where he spread Marxism–Leninism. He has been successfully employing strategic and soft power in undermining the West and America in particular. China is strengthening its armed forces and uses its financial assets to gain international political influence. It rents harbors in strategic locations all over the world. Americans are already hooked to some Chinese technologies owned by companies beholden to the Chinese Communist Party. Lian is convinced American and Western policy makers have underestimated the aspirations and successes of the Chinese Communist Party. "Western governments have been on to China's hard-power geopolitical-choke-point strategy," he says. "But they are only just beginning to appreciate its far more invasive soft-power strategy – and it might already be too late."[27]

Lian's view is one-sided and flawed in my opinion; while China has ambitions that try to recreate its imperial past, they hardly match those of America. The United States has been a world hegemon for decades. According to Jeffrey Sachs, professor of economics at Columbia University, the "U.S. security state still pursues a grand strategy of 'primacy,' that is, the aspiration of the U.S. to be the dominant economic, financial, technological, and military power in every region of the world."[28] China is copying America's playbook with considerable success. China has learned its American lessons well. It has imperial traditions, which it seeks to update and incorporate into its domestic and foreign policies.

An American Chinese reporter, Alice Su, says that Xi Jinping has been crowning himself with imperial power. He is a "disciplinarian," but not a revolutionary. Xi is imitating the legal philosopher Han Feizi who advised Qin Shi Huang, the first emperor of China. Han convinced Qin that people need law and punishment for crimes they commit, otherwise society becomes ungovernable. Xi is drawing from this legal, philosophical, nationalist, and imperial tradition for his vision of China. But he is

[27]Yi-Zheng Lian, "Trump Is Wrong About Tik Tok. China's Plans Are Much More Sinister," *New York Times*, September 17, 2020.
[28]Jeffrey Sachs, "The Perils and Promise of the Emerging Multipolar World," *Common Dreams*, June 6, 2024.

also a Marxist and a Maoist. His version of communism incorporates "Confucius and e-commerce." Xi Jinping "sees himself as a savior to lead the country into a new era of greatness propelled by rising prosperity and political devotion."[29]

China did so well in restraining the pandemic of 2020 that the country opened for business – and much more. In the March 2021 People's Congress in Beijing, the communist delegates were exuberant in praising President Xi Jinping and China's ambitious agenda to surpass the United States. The ambition is to develop more advanced technologies in artificial intelligence, quantum computing, integrated circuits, and aerospace. President Jinping was right, however, that everything a state does depends on self-confidence. The victory over the coronavirus plague, Xi said, resulted from "self-confidence in our path, self-confidence in our theories, self-confidence in our system, [and] self-confidence in our culture. Our national system can concentrate force [sic.] to do big things."[30] It is this complicated political context that explains, to some degree, agriculture, and ecological civilization in China. Agriculture in China has been trying to catch up with the agricultural superpower illusion of America. Yes, America produces huge amounts of food, but at unsustainable and catastrophic costs and consequences.

The Small Family Farmers of China

China has a growing sector of American-like agriculture. It also has about 260 million small family farmers, however. These rural smallholders have been practicing traditional farming for centuries. Caught between these two gigantic forces (American-style agriculture and traditional farming), the Chinese government is often seen campaigning on behalf of environmental and public health protection. The mantra of the Chinese government is ecological civilization.

The idea of ecological civilization is political and ecological. It may not be a façade. It may or may not be utopian. Ecological civilization is beautiful; it brings to mind the image of heaven on Earth: flourishing

---

[29]Alice Su, "The Rise of Emperor Xi: Prosperity, Power and Political Devotion Merge," *Los Angeles Times*, October 31, 2020.
[30]Alice Su, "China Delegates OK Plan to Surpass U.S.," *Los Angeles Times*, March 12, 2021.

villages and towns, farmers working the land without outsiders oppressing them or oppressing each other or polluting the natural world; flowers, monarch butterflies, honeybees, singing birds, sheep and lambs, fig trees, flowering lemon and almond trees, creeks and rivers running through the land, olive groves, grapevines, and the god Dionysos and his maenad followers indulging in a frenzy of dance and music. This picture of Chinese and Greek affections for the natural world does not necessarily incorporate political realities like those of the Mao and Xi eras. And neither does it mirror nature and society under the stress of the COVID-19 pandemic.

On July 1, 1958, Mao Zedong said farewell to the god of plagues.[31] Mao's celebration was short lived. The god of plagues has often been ravaging China and the rest of the world. China, in fact, started the twenty-first century with a plague. According to Yanzhong Huang, professor of diplomacy at Seton Hall University, the severe acute respiratory syndrome (SARS) plague of 2002 caught the government of China unprepared. This was as much a blow to public health as it was a political embarrassment to the Communist Party.[32] The 2020 pandemic was a warning. We must become sensitive to the nature, harmony, and extreme vulnerability of the world. Humans are one of millions of species. We have no right to procreate so fast and to exploit the natural world so intensely that we wipe out wildlife. Overpopulation is not appropriate for our times of climate emergency. Humans cannot and should not crush wildlife and expect immunity from such violence.

According to Aristotle, nature and animals are beautiful and perfect.[33] Chinese Dao thinking goes a step further. It considers humans as part of nature. We are one.[34] Nature is not dead and divorced from us. There are repercussions to everything we do in the natural world. We are connected to all animals, plants, waters, and the stars.

Zhihe Wang and Meijun Fan, who direct the Institute for Postmodern Development of China in Claremont, California, probably have their own dreams for ecological civilization. They grew up in the China of Mao Zedong. They experienced hunger and plagues, and witnessed the

---

[31] *Selected Works of Mao Tse-tung* (Maoist Documentation Project, 2007).

[32] Yanzhong Huang, "The SARS Epidemic and its Aftermath in China: A Political Perspective," in *Learning from SARS: Preparing for the Next Disease Outbreak* (Washington, DC: The National Academies Press, 2004) 116–136.

[33] Aristotle, *Parts of Animals* 645a15–36.

[34] Zhuangzi in Wikipedia. Both Greek and Chinese natural philosophers.

destruction of traditional Chinese culture. Nevertheless, they are messengers of the Dao philosophy. They are caught between China and America, Mao Zedong, communism, Xi Jinping, and extreme capitalism. They may see all these merging into ecological civilization as an emerging new global philosophy. Either humans will learn how to live in harmony with the natural world or they will become extinct.[35] Perhaps ecological civilization is a convenient expression for the end of strife and a beginning of something better for themselves and China. It may be no more than a slogan or deep belief in a better world.

I joined the discussion about ecological civilization during some of the conferences Wang and Fan sponsored in Claremont, California. That gave me a chance to talk to Chinese scholars. Such theoretical perspectives enriched my limited observations in rural China. I already mentioned that small Chinese farmers are passionately attached to the land they rent from the state.[36] And Chinese agronomists who study traditional farming told me they would love to see a better future for such farming. Yet the Chinese government and the American-trained scientists are confused about their own rural citizens. They overlook that these small family farmers are raising most of China's food, while the Chinese government supports the conversion of farmlands to large factory farms.[37] Such a policy is bound to spark clashes between rural citizens and large industrialized farmers supported by the government. This looming tragedy is a telling example of how difficult it is to maintain ancient ecological traditions in an age of worldwide ecocide and rapacious ambitions and governance.

In contrast to the hegemonic American agribusiness and the equally hegemonic if misguided developing Chinese agribusiness, Chinese peasant farming opens an exciting vista of ecological and political insights. Small family farmers favor a strategy of an agriculture that is largely benign to the natural world, while it benefits them who cultivate the land and grow healthy food for all. Industrialized agribusiness, be that of the American or Chinese variety, is against ecological civilization.

---

[35] Evaggelos Vallianatos, "China: The Other Side of the World," *Huffington Post*, October 16, 2017.

[36] Evaggelos Vallianatos, "Deep Passion for the Land – in China," *Truthout*, November 13, 2014.

[37] Ye Jingzhong, "China Encountering Modern Agriculture: Farming Without Traditional Farmers.?" In *Report of the International Symposium on Agroecology in China*, Kunming, Yunnan, China, 29–31 August 2016 (Rome: UN Food and Agriculture Organization, 2017).

Organic/biological farming and small family agriculture open the doors to ecological civilization – just a little. They give us but a glimpse of what the future could become. The first system – agribusiness – is about control and power; the second is a spark from millennial traditions of wisdom and practice in the raising of food without wounding the land. There's also the best of modern science coming under the name of agroecology: the latest findings in agricultural ecology, that could and would complement and enrich traditional practices. China will do its culture a favor if it turns all its efforts toward repairing and strengthening traditional farming and abandoning agribusiness. Such a policy shift would tell the world China is serious about ecological civilization and fighting global warming. At that moment, China might become a model for saving the Earth.

Ecological Civilization in China

In late-October to early-November 2019, the Chinese Academy of Social Sciences invited me to an international ecological conference in Jinan, Shandong Province. The Academy gave the conference a provocative and insightful title: "Paradigm Shift: Towards Ecological Civilization: China and the World." I listened to several Chinese and non-Chinese experts talk about a variety of issues (political, economic and ecological) touching on our present world crisis. The discussion took place during the last two days of October 2019. Chinese speakers had reasons for being exuberant. They merged their ecological dreams with their celebration of the 70th anniversary of the Chinese Revolution. Chinese forum speakers stressed their ideological victories of having institutions dedicated to the exploration of ecological civilization in all its complexity. In such pioneering tasks, they have the blessings of Xi Jinping, president of China – at least that is what I heard from my Chinese colleagues. A section of the forum examined "the world significance of Xi Jinping's thoughts on ecological civilization and Chinese traditional ecological wisdom."

Western participants like me brought out the looming threats industrialized civilization poses to human health and the health and very survival of the natural world. The picture that emerged was by no means pretty. Politicians, scholars and scientists spoke, sometimes passionately, about how to make China and the world better places, especially how to avoid the worst effects of climate change. I focused my remarks on the so-called industrialized agriculture. I tried to convey the fact that making farming a mechanical factory was no less a grave error than becoming addicted to petroleum, natural gas, and coal.

We have been undermining our health and the health and survival of the natural world. Here is how it happens: America, Europe, China and the affluent classes of most other nations have embraced giant farms growing a few selected crops. These large pieces of land are the present version of medieval plantations and twentieth-century state farms. Their corporate, state or private owners manage these farms like factories. They employ machines, genetic engineering for the modification of crops, and neurotoxic pesticides. The toxic cover of such large agricultural territories and the crops themselves are fatal to pollinating honeybees, other insects, birds and wildlife.[38] Poisons sip into the land and devastate microorganisms responsible for carrying nutrients to the crops. In addition, sprayed neurotoxins become airborne and travel with the winds. They contaminate the environment, including organic farms. The conversion of forests to industrial farms and the concentration of thousands of animals in gigantic animal factories make a substantial contribution to greenhouse gases warming the planet. I urged China to take the initiative in sponsoring a World Environment Organization for collective international activities for the transition of the world economy away from fossil fuels. Such actions and policies must be compatible with efforts to mitigate the emergency of climate change and over-industrialization of farms and food production.

Growing Agroecology in China

The second part of my visit to Jinan was eye-opening. I spent a day visiting a distinguished Chinese scientist by the name of Jiang Gaoming. He works in the Hongyi Organic Farm; his land in the village that gave him birth.

A Dutch colleague, Harris Tiddens, and I went from Jinan to Qufu, the hometown of the ancient Chinese philosopher, Confucius, who flourished in late sixth and early fifth centuries BCE. From Qufu we traveled to the Jiang Family Village located in Pingyi County, Linyi City. Jiang Gaoming is a man of knowledge and passionate about organic food and ecological civilization. He is associated with the Institute of Botany of the Chinese

---

[38]Evaggelos Vallianatos with McKay Jenkins, *Poison Spring: The Secret History of Pollution and the EPA* (New York: Bloomsbury Press, 2014).

Fig. 3.   Jiang Gaoming showing off his vegetables at Hongyi Organic Farm, Jiang Family Village, Pingyi County, Linyi City, Shandong Province, China. Photo by Evaggelos Vallianatos.

Academy of Sciences and the Shandong Province that funds his research. He is a prolific botanist interested in public health and the health of the natural world. He grows organic food and tests plants for their food and medicinal virtues.

Jiang Gaoming, his two graduate students, the farm manager, Harris Tiddens, Gao Yuan, a graduate student in the philosophy of science at Beijing Normal University, and I sat around a round wooden table for dinner. On it were five bowls containing delicious vegetables, noodles, and rice. Each of us had a pair of wooden chopsticks for taking food from the bowls. In addition, Jiang Gaoming kept filling our tiny glasses with a drink from sorghum and sweet wine. This memorable symposium led to extensive talk. I listened to him describing his work and marveled at the breadth of interest and deep knowledge he possessed. He is a professor of plant ecological physiology. In other words, he is investigating the natural history of plants that make life possible. Ecology is his mission. He and his graduate students are paving the path for China to enter the scientific and political realms of ecological civilization. For the next half of the day, Jiang gave us a tour of the various strips of land where he and his graduate students are testing plants. His German Shepherd dog, Tiger, followed us everywhere. We even went to the center of his village where a small store holds his books for sale. I departed China with the botany professor in mind.

Ecological Civilization, Climate Change, and Pandemics

Talk about ecological civilization is sweet. I don't think that anyone knows what ecological civilization was, is, or if it is possible among humans. But we know traditional Greek and Chinese wisdom and institutions are the closest possible models of ecological civilization. Yet it is great to have gigantic dreams of one day converting humans hooked on petroleum and pollution to caring for the Earth like the ancient Greeks and ancient Chinese did. And then the 2020 pandemic complicated things. In my opinion, it seems likely to have originated in China. I will explain more about it later in this book. The pandemic nearly stopped the world. I will be arguing that agribusiness has been at the center of the pandemic.

Xi Jinping ought to start the conversation about climate and ecological civilization with our hospitable, friendly, and ingenious Professor Jiang Gaoming. He is growing agroecology, which promises to be a new species of ecological civilization.

The next chapter is about the economic and political effects of giant farms: how and why industrialized agriculture has been wrecking family farms and democracy.

3 Big Ag: More Harm than Good?

Cows at a small animal farm in Central Valley, California.

Photo by Evaggelos Vallianatos.

American Enclosures

In 1950, there were enough farmers around to account for 15.2% of the population. By 1969 that number had dwindled to 5.1%. And by 1978, all the land raising crops or cattle was owned by 7.7% of the families in America. From these families, only 62,260 owned 3 out of every 10 acres of farm and ranch land. In 2002, America had about 2 million farms, but 3.7% of those farms, each cultivating more than 2000 acres of land, produced most of the country's food. Only 67.1% of America's farmers or 1,428,136, owned their land; about 25.9% (551,004) were part owners; and 7.0% (149,842) were tenants.

The decline of the people at the farms has been followed by a spectacular rise and multiplication of machines: The tractor, and other petroleum-powered machines, and myriad chemicals did away with the peasant and family farmer. In 1920, the United States had a total of 5 million tractor horsepower and 13,406 million man-hours spent on farms. In 1950, farmers used 93 million tractor horsepower and 6922 million man-hours to cultivate their farms, while spending US$5,640,000,000 to operate their machines. In 1969, this small agricultural class had at its command an entire mechanical army of 203 million tractor horsepower whose running cost alone were US$11,500,000,000 dollars. The growers of 1969 also made use of 3431 million man-hours, most of which were cheap labor extracted from migrant farm workers.[1] Agribusiness replaced family farming with chemical factory agriculture. This explains why the Environmental Protection Agency (EPA) has been running a large bureaucracy for the regulation/approval of pesticides for the chemical factories producing food. The owners of those factories (sometimes known as farmers, growers, and ranchers) think nothing of sowing seeds which the

---

[1] My numbers for these changes in the US food system come from unpublished EPA data. However, United States Department of Agriculture (USDA) agricultural statistics illustrate the trends in the diet, food, farm population and agricultural practices in America. Of the scholars who have studied agriculture, David Pimentel of Cornell University provides excellent data and insights: D. Pimentel and M. Pimentel, eds., *Food, Energy and Society*, 3rd edition (Boca Raton, FL: CRC Press, 2008); David Pimentel *et al.* Reducing energy inputs in the U.S. food system, *Human Ecology*, **36** (4), (2008): 459–471; Henry W. Kindall and David Pimentel, Constraints on the expansion of the global food supply, *Ambio*, **23** (3), May 1994. See also: Evaggelos Vallianatos, *This Land is Their Land: How Corporate Farms Threaten the World* (Monroe, Maine: Common Courage Press, 2006), 137–196.

chemical industry has "dressed" with all sorts of neurotoxic and carcinogenic poisons, especially fungi-killing ones.[2]

Land Grant Universities: The Brains of Agribusiness

My 25-years of experience at the US EPA convinced me that pesticides mirror corrupt science funded by the owners of giant industrialized agriculture. Such a combination has been killing nature while giving power to agribusiness in the United States. Millions of family farmers used to have a stake in the agrarian economy in rural America, but only four companies had monopoly power in each sector of the food and agricultural system by the end of the twentieth century. For example, in 1998, IBP, ConAgra Beef, Excel Corporation (Cargill), and Farmland National Beef Packing Company slaughtered 79% of cattle in the United States. Six companies, Smithfield, IBP, ConAgra (Swift), Cargill (Excel), Farmland Industries and Hormel Foods, slaughtered 75% of all pigs in the country in 1999. Four companies, Cargill (Nutrena), Purina Mills (Koch Industries), Central Soya and Consolidated Nutrition (ADM and AGP), controlled all feed plants in the United States by 1994. Also, by 1997, four companies, Cargill, ADM Milling, Continental Grain and Bunge, controlled America's grain trade: they managed 24% of the grains, 39% of the facilities for storing grain, and 59% of the grain export facilities in the country.[3]

Meanwhile, some scientists are on the margins of this great consolidation of power in rural America. This process is not kind to both farmers and the natural world. Agribusinessmen manage mechanical and chemical empires that produce machineries and pesticides to replace traditional

[2]Rachel Carson, *Silent Spring* (Boston: Houghton Mifflin, 1962); Carol Van Strum, *A Bitter Fog: Herbicides and Human Rights* (San Francisco: Sierra Club Books, 1981); Rober Van Den Bosch, *Pesticide Conspiracy* (Berkeley: University of California Press, 1989); Evaggelos Vallianatos with McKay Jenkins, *Poison Spring: The Secret History of Pollution and the EPA* (New York: Bloomsbury Press, 2014).

[3]US Department of Agriculture, *Concentration in Agriculture: A Report of the USDA Advisory Committee on Agricultural Concentration* (Washington, DC., June 1996); William Heffernan, *Consolidation in the Food and Agriculture System* (Report to the National Farmers Union, February 5, 1999).

farming methods.[4] They also evict small family farmers from the land.[5] Other scientists, especially those of the land grant universities, have been legitimizing the horror of changing the society of rural America from small family farmers to food factories. Land grant universities were founded by the Morrill Act of 1862.[6] Congressman Justin Smith Morrill of Vermont introduced the land grant college bill and President Abraham Lincoln signed it. Morrill and Lincoln believed that great innovation would help family farmers. They had been funded by federal and state governments and industry. As far back as 1980, Don Paarlberg, a senior official of the US Department of Agriculture (USDA), admitted that the land grant colleges failed their mission; boosting the industrialization of agriculture instead of serving family farmers. "The Extension Service," he said, "with its advice that a farmer should have a business 'big enough to be efficient,' undoubtedly speeded up [sic.] the process of farm consolidation and reduced the number of farms. In the classroom, emphasis on modern management helped put the traditional family farm into a state of total eclipse."[7] Paarlberg was right. The 76 land grant universities are different from the original agricultural colleges. They have become the brains of agribusiness, thinking and inventing all the gadgetry, machinery and chemicals fueling America's gigantic farms and agribusiness.[8]

Land grant universities designed animal farms. It does not bother them that it is wrong to treat animals like inanimate things good only for eating. It is as if the beauty, history, and extraordinary value of family farming, and nature as sources of life, democracy, and culture (what the Greeks would call Mother Earth) disappear whenever someone throws money or power at scientists; or when the state adopts nature-killing

---

[4]Robert Van Den Bosch, *The Pesticide Conspiracy; Christopher Ketcham, How Cowboys, Capitalism, and Corruption are Ruining the American West* (New York: Viking, 2019).

[5]A. V. Krebs, *The Corporate Reapers: The Book of Agribusiness* (Washington, DC: Essential Books, 1992); US Department of Agriculture, *A Time to Choose: Summary Report on the Structure of Agriculture* (Washington, DC: USDA, January 1981).

[6]Genevieve R. Croft, *The U.S. Land-Grant University System: An Overview*, Washington, DC: Congressional Research Service, August 29, 2019.

[7]Don Paarlberg, The land grant colleges and the structure issue, in *A Time to Choose: Summary Report on the Structure of Agriculture* (Washington, DC: US Department of Agriculture, January 1981), p. 129.

[8]Food and Water Watch, Public research, private gain: Corporate influence over agricultural research, April 2012.

policies – in this case, subsidizing the takeover of rural America by corporations and large farmers. Then nearly all debate becomes muted.

I taught at the University of Maryland for a year-and-a-half, from 2002 to 2004. The University of Maryland is a land grant school. Most of my former 40 colleagues in the Department of Natural Resource Sciences, and not merely them, had nothing to do with family farming. Only two tenured professors offered courses relevant to organic farming from time to time. The rest refused to use the words "sustainable farming," "organic agriculture," or "family farming." Their technical discourse was empty of democratic concern for the land and public health. It was all about ecosystems management and control, pest control, and integrated pest management. They kept teaching and researching "nutrient" management as if they were trying to hide the ceaseless suffocation of the Chesapeake Bay, Maryland's water treasure, and one of the country's ecological, fishing, and recreational jewels, by the wastes of the chicken factory farms of Delmarva, comprising of Delaware, Maryland and Virginia. Individual chicken farms in Maryland in 2008 could host as many as 20,000 to 30,000 chicken. All together, in 2008, Maryland had a population of 570 million chickens whose manure accounted for 650 million pounds. With these overwhelming numbers of chicken, pollution was severe. Blue crabs and oysters suffered in the Chesapeake Bay.[9] Pollution comes to Chesapeake Bay from as far away as Pennsylvania, Virginia, Delaware and Maryland.[10] The real danger lurks next door in the animal farms, which are all but invisible at the University of Maryland.

Professors in the country's 68 land grant universities value pesticides but appear to fear family farming would brand them as radicals or liberals. But the real reason for self-censorship and immersion in technics mirrors their commitment – in the case of Maryland – to chicken agribusiness. And in the case of the rest of the land grant universities, to the mission of agribusiness: the dazzling new prospects of biotechnology; training students to manipulate life with genetic engineering and molecular biology; feeding all students fast food; teaching the few students studying agricultural science and natural resources extremely narrow technical skills about soils, water, crops, forests, insects, plants, fish, and wildlife,

---

[9] Ian Urbina, In Maryland, focus on poultry industry pollution, *New York Times*, November 28, 2008.

[10] US EPA, Office of Wetlands, Oceans, and Watersheds, *Guidance for Federal Land Management in the Chesapeake Bay Watershed, Agriculture*, May 12, 2010.

and making the factory farms palatable and inevitable, being products of science.

Some academics are teaching politics, natural science, and social science. However, they divorce themselves from society. They probably feel indifferent or comfortable that machines, large farmers and industrialists are "developing" rural America.[11] They exalt the successes of some family farmers who, following the path of organic farmers, resist corporate farming and extinction. But these daring family farmers, including organic farmers, are not many in number. In 2005, Maryland had 5.5 million people, of which about 12,000 were growers; 7680 of whom in 2000 received subsidies, while the rest – the "hobby" farmers – were too impoverished to qualify for anything. Maryland in 2007 had only 89 organic farmers. Ten years later, in 2017, Maryland had 117 organic farmers.[12] More than 35% of farm receipts in Maryland come from chicken, produced like so many pieces of machinery: In 2003, Maryland growers produced 292 million broiler chicken.[13] The chicken factories of the Delmarva Peninsula, which includes those of Maryland, leave a vast footprint of ecological devastation, social upheaval, and widespread disease among animals, humans, and the natural world.

Animal Outlook,[14] an animal welfare organization, described America's animal farm tragedy this way: "Workers on farms and processing plants, a workforce with many vulnerable undocumented workers, often endure dangerous conditions, long hours, and low pay. People begin to suffer along with the birds." Animal suffering includes "torturous [living] conditions" and barbaric practices. Pollution from animal farms and crop production threatens more than 13,000 miles of rivers and streams, and over 60,000 acres of lakes and ponds. In Wicomico County, Maryland, the County Health Department reported, "one in four middle school students diagnosed with asthma... Communities all along the east coast – like towns in Delaware and North Carolina – are suing meat companies for offenses from environmental nuisance to wrongful death."

---

[11] Bosch, *The Pesticide Conspiracy*, 119–128.

[12] US Department of Agriculture, Maryland Agriculture Has it All, June 4, 2019.

[13] Maryland Agricultural Statistics, *Poultry Review*, 2003.

[14] Animal Outlook, Factory farms make animals, people and the environment sick, September 20, 2018.

Destabilizing and Harmful Economic and Political Effects of Corporate Farms

The social consequences of the animal factories that produce America's meat are harmful and of long standing. From 1995 to 1996, animal farms had a population of 103 million pigs, 58 million non-dairy cattle, 7.6 billion chickens, and 300 million turkeys. From 1969 to 1992 the number of chicken farms declined by 35%, yet the number of chickens *produced* almost tripled. A mere 2% of the non-dairy cattle farms sell more than 40% of the cattle for slaughter. And since 1982 a drastic concentration of power has been taking place in the pig farms – their overall numbers were reduced from 600,000 in 1982 to 157,000 in 1997.[15]

A few Americans, including a rare politician or two, have been critical of this power grab and dramatic decline of small family farmers. On April 16, 1968, the Governor of North Dakota, William L. Guy, wrote a letter to Senator Gaylord Nelson in which he explained why he detested agribusiness. "It should not take a very sophisticated study," he said, "to show that large corporation farming eliminates the need for small farm units living on the land. When small farm units are eliminated and the families who farmed them are moved to the cities, some very grave economic and social problems arise in the rural areas. Personal property tax income diminishes; and the financial support and, for that matter, the need for such things as schools, churches, recreation and health facilities in rural communities diminish... North Dakota has barred corporation farming since the middle 1930s. During those depression years, land foreclosures placed so many farms in the hands of the corporate lender that there was grave danger of the majority of our state's farmland being in the hands of corporations... I grew up in the shadow of the corporation farm in North Dakota... This corporation owned its own grain marketing facilities in several adjacent towns. It bought its machinery direct from the Minneapolis-Moline farm equipment manufacturing company. It formed a cooperative gasoline and oil company which granted credit and gave service to the farm tenants of the corporation. The retail gasoline dealer was forced out of business. At one time, this corporation owned every building in the corporation town. Businesses were run on a concession basis. It might be argued that this could never come to pass again,

---

[15] *Animal Waste Pollution in America*, p. 3.

but I believe that it could... I am strongly opposed to corporations taking over the farming industry."[16]

Few American politicians understood the danger of agribusiness as Governor Guy did. Senator Nelson from Wisconsin understood the danger. He spent the 1960s and 1970s holding hearings about corporate power. He, like Governor Guy, opposed the takeover of rural America by factory farms. His hearings forced the country (and particularly the leadership of the country) to listen to voices – many, eloquent, and passionately defending family farming – it did not wish to hear. John Helmuth, an economist who teaches at Iowa State University, saw the rotten core of animal factories quite early. He said in 1990, "The meatpacking industry has rapidly become an industry where literally every day, fewer and fewer individuals are making more and more of the economic decisions about what America eats... this consolidation of economic power in the meatpacking industry has resulted in (and is resulting into a greater degree every day): lower prices paid to farmers and livestock producers, lower wages and deplorable working conditions for meat industry workers, serious questions about the quality and nutritional safety of meat, and higher prices paid by consumers. When an industry drives its best small and medium companies into bankruptcy, when cattle producers and farmers are driven into bankruptcy by lower and lower prices, when workers are treated like animals and injured and maimed for life, and consumers are charged higher and higher prices for minimum quality, often unsafe meat, something is wrong."[17]

Helmuth illustrates the nature of that wrong by highlighting the "breathtakingly rapid" consolidation and concentration of economic power in the slaughter or meatpacking industry, particularly in the 1980s. "Never in any American industry in any other time period," he says, "has there been such a huge and rapid seizure of economic power."[18] For example: In 1998, four companies – Iowa Beef Processors, ConAgra Beef

---

[16]Letter of Governor William L. Guy to Senator Gaylord Nelson in "Role of Giant Corporations," Hearings Before the Subcommittee on Monopoly of the Select Committee on Small Business, United States Senate, Ninety-Second Congress, First and Second Sessions, Part 3A, Corporate Secrecy: Agribusiness, November 23 and December 1, 1971; March 1 and 2, 1972 (Washington, DC: US Government Printing Office, 1973), pp. 4179–4180.

[17]John W. Helmuth, Introduction, in Krebs, *Heading Toward the Last Roundup*, p. ix.

[18]Helmuth in Krebs, p. vi.

Companies, Excel Corporation (Cargill) and Farmland National Beef Pkg. Company – controlled 79% of beef slaughter in the United States; Smithfield, Iowa Beef Processors, ConAgra (Swift), and Cargill (Excel) slaughtered 44% of all pigs in 1992; Tyson Foods, Gold Kist, Perdue Farms and Pilgrim's Pride slaughtered 49% of all chicken in 1998; Cargill, ADM, Continental Grain, and Bunge control America's grain trade: their grain elevators hold 24% of the country's grain, and own or control 39% of the grain facilities and 59% of the port facilities for grain trade. Finally, ADM Milling Company, Corn Agra, Cargill Food Flour Milling and Cereal Food Processors controlled 61% of flour milling business in 1990.[19]

Chuck Hassebrook, director of the Center for Rural Affairs in Nebraska, was equally passionate in support of family farming and small businesses. He said in November 2001 that since 1996, giant hog owners have driven half of the family hog farmers out of business. That way, corporate agriculture replaces "genuine middle class self-employment opportunities in family farming with low-wage jobs that most rural people won't take...Mega livestock production will not offer lasting growth. Big hog production is already shifting to Mexico and Brazil...The animal science departments of land grant colleges, with some exceptions, have not stepped up to help us chart a better course. Too often, they have acted as cheerleaders for industrial livestock production rather than public institutions with a responsibility to empower rural people to create a future that reflects their values."[20] This extreme concentration of economic and political power in the hands of relatively few corporate executives undermines rural America – such that, it is practically a rule that wherever big agriculture settles, it is almost certain it will rapidly destroy the democratic, cultural, and economic institutions of the community, and convert everything around it to a plantation.

---

[19]William Heffernan, Robert Gronski and Mary Hendrickson, *Concentration of Agricultural Markets – January 1999* (Department of Rural Sociology, University of Missouri, Columbia, Missouri, 1999). See also William Heffernan, Confidence and courage in the next 50 years, *Rural Sociology*, 1989, **54** (2), 149–168; and Agriculture and Monopoly Capital, *Monthly Review*, July/August 1998, pp. 46–59.

[20]Chuck Hassebrook, Rural-friendly livestock industry model needed, *Center for Rural Affairs Newsletter*, November 2001, p. 8; See also: Editorial, The curse of factory farms, *The New York Times*, August 30, 2002.

Tony T. DeChant, president of the National Farmers Union, testified in Congress on May 20, 1968, that corporate farms "represent power without [a] conscience."[21] Ben H. Radcliffe, president of the South Dakota Farmers Union, also testified before Congress on May 20, 1968. He told the Senators that social and economic decay follows large farms all over the country; in his words, "You can drive almost anywhere in the rural areas, and see the results of our failure to weigh social consequences in determining our economic objectives: the weathered, abandoned farmhouse, a curtain flapping through a broken window; the soaped-up plate glass of the store front with the 'closed' sign taped to the door; the weeds standing tall around the vacant service station, and the growing ratio of older people on our main streets in areas like South Dakota."[22]

Arvin and Dinuba in the 1940s: Decay of the American Character

This decay is not confined to South Dakota. For example, Arvin is a small rural town in southeastern Kern County in the fertile Central Valley of California. In 1940, factory-like farms surrounded Arvin. Arvin, however, did not share in the prosperity of those factory-like farms. The average farm size of Arvin was 500 acres. Only 35% of Arvin farmers owned their land. Of the local population, only 4% of the people of Arvin were native Californians; 63% were Dust Bowl migrants with less than five years of residence in the town. They earned little and did not have much interest in their community. Even the managers of the large farms were absentees. If any businessmen lived in Arvin, they went to Bakersfield and Los Angeles for recreation. Arvin's schoolteachers found the town so distressing that most of them lived in Bakersfield, commuting 22 miles daily. The elementary schools, churches, and the economy of Arvin were impoverished. Arvin had no high school. The town had no elected political leadership of its own. It was unincorporated. Its large farms converted it into a slum and a colony. We know these things about Arvin because of Walter Goldschmidt, an anthropologist with the US Department of Agriculture in

---

[21] Statement of Tony T. DeChant, president, National Farmers Union, Denver, Colo. in *Role of Giant Corporations*, Part 3A, p. 4181.

[22] Statement of Ben H. Radcliffe, president, South Dakota Farmers Union, Huron, S. Dak., *Ibid.*, p. 4193.

the early 1940s. Goldschmidt (cited below) studied Arvin and compared his findings with what he discovered in another rural town that was dominated by small farms. That rural town was Dinuba in northern Tulare County in California's Central Valley.

In 1940, the average size of farms in Dinuba was 57 acres. More than 75% of the farmers of Dinuba owned their land. Dinuba's economy and culture were vigorous and democratic. Its elementary and high schools were good. The teachers lived in town and made outstanding contributions to the culture of the community. Dinuba's residents were middle class persons with good incomes and a strong interest in their town. 19% of the people of Dinuba were native Californians and 22% Dust Bowl migrants. The median length of residence at Dinuba was between 15 and 20 years. Dinuba's prosperity was the prosperity of its small farms. Yet Dinuba and Arvin were similar rural towns. They enjoyed the same climate and fertile land. They were equidistant from small and large cities and had access to highways and railroads. They had the same industrialized farming, relying on laborers to do the hard and dangerous work. They specialized in single crops, which they produced exclusively for cash sales. Dinuba raised fruits, especially raisin grapes, some cotton, and vegetables. Arvin produced largely cotton, potatoes, fruits and vegetables, grapes, and grain.

The sole factor that made Arvin and Dinuba different was the size of the farms – Arvin had large farms and Dinuba had small farms. 91% of Arvin's land was in farms larger than 160 acres but only 25% of Dinuba's land was in farms of over 160 acres. The economic, social and democratic consequences of farm size in Dinuba and Arvin were dramatic: Dinuba's small-farm economy supported 62 business; Arvin's large-farm economy: 35. The volume of retail trade in Dinuba for a year was US$4,383,000 while Arvin's amounted to US$2,535,000. The small-farm community spent over three times more money on household supplies and building equipment than the large-farm community. More than one-half of the breadwinners of Dinuba, but less than one-fifth of the breadwinners of Arvin, were independent businessmen, white-color workers or farmers. Less than one-third of the breadwinners of the small-farm community were agricultural workers while nearly two-thirds of those gainfully employed in the large-farm community were agricultural workers. Dinuba had three parks and two newspapers. The town had paved streets, sewerage, and streetlights. Arvin had but a single playground loaned by a corporation and one newspaper. Arvin had practically no paving, streetlights, or sidewalks. It had inadequate water and sewage facilities. For these reasons, Goldschmidt said, Arvin was "less a community than an agglomeration of houses."

Goldschmidt was right. Arvin was not merely a disintegrating rural community in California's Central Valley but the nightmare of rural America. Agribusiness badly impacts family farming and industrializes the countryside. Even the government, particularly the USDA, was becoming a subsidiary of agribusiness. This was bad enough, and not merely because of the concentration of land and power at the hands of a few men. Goldschmidt accused agribusiness of destroying the "American character" which, he said, "was forged in its rural hinterland: the frontiersman melding into the freeholding farmer created a pattern consisting of egalitarianism, personal independence, the demand for hard work and ingenuity, self-discipline, with the ultimate reward in a personal success." Despite the national importance of the freeholding farmers, however, the United States abandoned them to agribusiness, which either kicks them off the land or, Goldschmidt says, remakes them into "organization men in overalls."[23]

Both Governor Guy and Senator Nelson were aware of Goldschmidt's 1944 study on the adverse effects of huge farms on small communities in California's Central Valley, probably the richest land in the United States. They also knew of Goldschmidt's troubles with his superiors at USDA who were very unhappy with him. Goldschmidt had stirred up a hornet's nest and the large landlords of California were displeased. He wisely gave his report to Senator James E. Murray who was Chairman of the Senate Special Committee to Study Problems of American Small Business. Senator Murray appreciated the importance of Goldschmidt's investigation, so he published that study as a committee report on December 23, 1946.[24]

In the introduction to his study, Goldschmidt described the family farm as "the classic example of the American small business enterprise... the spread of the family farm over the land has laid the economic base for

---

[23]"Small Business and the Community: A Study in Central Valley of California on Effects of Scale of Farm Operations," *Report of the Special Committee to Study Problems of American Small Business*, United States Senate, Seventy-Ninth Congress, Second Session, December 23, 1946 (Washington, DC: US Government Printing Office, 1946).

[24]Small Business and the Community: A Study in Central Valley of California on Effects of Scale of Farm Operations (1946).

the liberties and the democratic institutions which this Nation counts as its greatest asset." Goldschmidt spoke like the Greek fourth-century BCE thinker, Xenophon, who equated family farming to freedom and civilization.[25] Goldschmidt then explained why Dinuba was part of the American tradition of agrarianism and why Arvin was a rising threat to that tradition. Even the businessmen of Arvin frequently expressed "their own feelings of impermanence; and their financial investment in the community, kept usually at a minimum, reflects the same view. Attitudes such as these are not conducive to stability and the rich kind of rural community life which is properly associated with the traditional family farm." [26] Goldschmidt's fear was an ancient fear. He discovered, in the fertile land of California, "too few farms and those too large." He knew that large farms were inimical to the health of democracy – there's simply no way to misread history at this crucial political evolution of societies. Political observers in California were right to say, "A Republic cannot long survive when the lands are concentrated in the hands of a few men."[27]

A Walnut Grower with a Cannon

In the summer of 2012, I visited a large walnut farmer in the neighborhood of Arvin and Dinuba, in the gigantic Central Valley of California. The farmer/grower had cancer but declared pesticides innocent of all harm. He told me that his doctor assured him, "pesticides had nothing to do" with his cancer. This farmer, an exceptionally kind and hospitable man, also thought nothing about global warming, dismissing it as a hoax. And when I urged him to slowly abandon his conventional agribusiness practices for organic farming methods, he laughed. He said, "Socrates, you are eating my food, and it is not organic. I use the minimum amount of water for my trees." (He probably called me Socrates because of the trouble he had in pronouncing Evaggelos.) He said his troubles are solely with regulators. He employs about 10 workers from Mexico but has no confidence that American citizens would be interested in working for him

---

[25] Xenophon, *Oeconomicus* 5.1–20.
[26] "Small Business and the Community" was reprinted in *Role of Giant Corporations*, Part 3A, pp. 4465–4648. Quotes from the introduction of Goldschmidt's study, pp. 4474–4477.
[27] *Transactions of California State Agricultural Society*, 1884.

or any other farmer. Behind his home, he had a workshop filled with machines: several tractors, a variety of trucks, and specialized machinery for the cultivation and harvesting of walnut trees. But the man also likes guns. His wife said there's a state prison not far away from her home. Her husband, she said, must be prepared for all eventualities. This meant owning lots of guns. Their armory even includes a replica Civil War cannon. The heavy cannon was on a truck platform. I asked him if he ever fired the cannon, and he immediately invited me to see him use the cannon. His wife and I got into his truck. He drove the truck, already hooked to the platform carrying the cannon, to the middle of his property. This was a country road next to a cement ditch. The countless walnut trees looked naked. Like soldiers in a geometrical pattern and without leaves, they hugged the evening air and declining light. The Sun looked red, painting the western sky with bright color. Meanwhile, the walnut grower climbed on the platform and slowly prepared the cannon, which he blasted twice. Instead of real cannon balls, he uses bowling balls. I watched the cannon carefully but saw no balls coming out of the cannon. A terrible sound and smoke covered the platform. Instinctively, I moved backwards. I looked at my smiling farmer friend with both disbelief and delight. His political views, like his cannon, are more in accord with those of Republican politicians, though he is liberal on religion.

A Small Animal Farm

My large farmer friend invited me on a tour of a family dairy holding some 200 cows. The animals looked healthy, and the dairy was clean. A dozen cows at a time went within iron enclosures where machines milked them. One did not see a drop of white milk, only tubes grabbing the teats of cows for a few minutes, sucking all the milk. A worker in tall black plastic boots walked next to the iron enclosures inspecting the milking machines. Then, suddenly, the enclosures opened and one cow after another, their ears labeled with large bright numbers, walked away relieved of their milk until the next period of milking. I looked at some of the cows and a few young calves in the eye, trying to get close to them. They backed off. Neither of us felt comfortable with the other. The owner of this family animal farm is by no means an agribusiness Cyclopes. However, he is married to federal subsidies and to schemes of trying to control the natural world. He is confident he is doing nothing wrong. He follows science and a political tradition.

World War II and the temptations of empire took the government out of the hair of the cowboy millionaires. They came up with another pernicious idea whereby they started concentrating thousands of animals under one roof. They lost no sleep over packing animals into cement enclosures like sardines in a can; giving them no chance to eat grass out in the open, breathe fresh air or see the light of the Sun. It did not occur to them that such treatment was extremely cruel and inhuman – the equivalent of treating the animals like pieces of machinery. In addition, this awful stress shatters the animals' immune defenses, making them sick. Add the ammonia and other toxic chemicals and pathogens in the air, and the imprisoned animals would be perpetually in the process of illness and dying.

The academic consultants from the agricultural universities who cooked this model of factory farming did not dare suggest that such a cruel practice was bound to erupt into disease for the animals, the poorly paid and often foreign workers feeding them, as well as the community around the animal camps; or that quite possibly, the diseases might evolve from endemics to become pandemics capable of spreading to the entire world.

Cowboys, however, could not be embarrassed. They make a living by slaughtering thousands and millions of animals. Ethical, philosophical, and health questions about their treatment of animals and the workers watching over the animals and doing the slaughtering and meat processing leave them cold. Massive lagoons hold tenuously enormous amounts of urine, feces, and a cocktail of drugs and poisons used in the animal farms. These lagoons often overflow into creeks and rivers, killing fish and poisoning the water.[28] In giant farms, even getting milk from a cow is not violence free. During his Academy Award speech, the actor Joaquin Phoenix said: "We feel entitled to artificially inseminate a cow and when she gives birth we steal her baby, even though her cries of anguish are unmistakable… And then we take her milk that's intended for the calf, and we put it in our coffee and cereal."[29]

---

[28]Robbin Marks, *Cesspools of Shame: How Factory Farm Lagoons and Sprayfields Threaten Environmental and Public Health* (Natural Resources Defense Council, July 2001); Rolf U. Halden and Kelogg J. Schwab, Environmental Impact of Industrial Farm Animal Production, *A Report of the Pew Commission on Industrial Farm Animal Production*, 2007.

[29]Quoted in Andrew Jacobs, Is dairy farming cruel to cows? *New York Times*, December 29, 2020.

Fig. 1.   Calf #1571, Central Valley, California. Photo by Evaggelos Vallianatos.

Animal factories cram millions of animals in thousands of small spaces, causing diseases, pain, and suffering of cruel proportions. "Milk-fed" veal comes from baby calves spending their short lives in the dark, chained to a wooden box so small the animal can't even turn around. However, the baby animal is mostly seen as a prized food, nothing else. In 1983, a reporter wrote: "calves less than a week old... are... reared in confinement so that they can be slaughtered and labeled "gourmet milk-fed veal," the epicurean delight that commands premium prices."[30] Once I saw a few-days-old calf in a farm at the Central Valley of California. It had tags with the number 1571 on both ears (Fig. 1). It turned and looked at me, his big eyes telling me of its horrible and inevitable fate: having been taken away from its mother probably immediately after its birth and expecting slaughter soon, so the farmer might sell veal.

The calf is made anemic in order to give its meat a special color. Erin Wing, former undercover investigator of animal abuse for an animal welfare organization, Animal Outlook, said she witnessed violence against cows and calves: "young calves have their horns brutally burned or gouged from their skulls in a cruel process known as 'dehorning' or 'disbudding.' Ultimately, the journey of a cow on a dairy farm is one of never-ending misery. If a cow manages to survive the torment of the

---

[30] Gail Ann Eisnitz, Is the veal being mistreated? *New York Times*, July 17, 1983.

Fig. 2. A white bull on Hongyi Organic Farm, Jiang Family Village, Pingyi County, Linyi City, Shandong Province, China. Photo by Evaggelos Vallianatos.

factory farm, she is forced and electrically prodded onto a truck bound for slaughter."[31]

Animals are living beings. They have feelings of enjoyment and fear. When I looked at the big eyes of the condemned calf, I felt desolate. It felt like the calf was talking to me, pleading for its life. Once, on a visit to China, I saw a white bull with absolute terror written in its eyes (see Fig. 2).

In order to keep calves and all the other factory animals alive, they drug them as much as they feed them. Farm animals continuously eat genetically engineered feed drenched with pesticides and antibiotics. The excrement and urine from these billions of farmed animals could potentially pollute the natural world. As we will see in the next section, they also make a substantial contribution to climate change.

The Animal Connection to Climate Change

Animal farms and industrial crop farms are responsible for something like 18 to 51% of CO_2 equivalent (CO_2-eq) of greenhouse gases per year. In 2006, the UN Food and Agriculture Organization (FAO) estimates the livestock sector was responsible for 18% of CO_2-eq of greenhouse gases;

---

[31] Erin Wing, Dairy is Dead on Arrival (Animal Outlook).

37% of methane emissions, which are 23 times more lasting than CO_2; 65% of nitrous oxide, which is 296 times more potent than CO_2; and 64% of ammonia, which makes acid rain. In 2009, two researchers from the World Bank Group criticized the FAO estimate. They calculated the total emissions of greenhouse gases by farm animals to be around 51% of CO_2-eq in greenhouse gases.[32]

In 2018, two British academic scientists[33] gave us the following math about the role of agriculture in warming the planet. They said that agriculture used about 43% of ice-free and desert-free land in the world as well as 90–95% of fresh drinking water. Of the gigantic amount of land employed for food production, about 87% was dedicated to raising food and 13% grew textile crops and crops for biofuels. Something like 550 million farms (ranging from the very small, to the medium, and very large), were feeding 7.6 billion people. That process of producing food for these billions of humans degrades land, waters, and society, and drives climate change. These scientists pointed to the price we pay by raising our food with machines burning petroleum and fueled by lots of chemicals. The "food supply," they said, was responsible for about 13.7 billion metric tons of CO_2-eq, which is 26% of the anthropogenic or man-made greenhouse gas emissions into the atmosphere. An additional 2.8 billion tons of CO_2-eq, or 5%, are generated from non-food agriculture and deforestation. Food production was also responsible for severely damaging the land of the planet: causing about 32% of acidification of the waters of lakes and rivers, which harms water animals. The other damage of the waters comes from the seasonal runoff of pesticide poisons and fertilizers. Food production causes about 78% of this deadly effect known as eutrophication: fertilizers from farms enter rivers and lakes and cause the fast growth of water plants. Lots of water plants growing fast deplete the oxygen in the water, which result in the asphyxiation of fish and other water animals, producing dead zones in rivers, lakes and seas. The Mississippi

---

[32] Henning Steinfeld *et al., Livestock's Long Shadow* (Rome: UN Food and Agriculture Organization, 2006); Robert Goodland and Jeff Anhang, Livestock and Climate Change: What if the key actors in climate change are cows, pigs, and chickens? *World Watch,* November/December 2009.

[33] J. Poore and T. Nemecek, Reducing food's environmental impacts through producers and consumers, *Science,* June 1, 2018.

River, for example, drains a great swath of the agricultural Midwest, giving rise to gigantic dead zones in the Gulf of Mexico.[34]

The consequences of these harmful practices and deleterious effects are the destabilization of ecosystems, the shrinking and degradation of biodiversity, and the invitation to disaster in the raising of food and the survival of humans and their civilization. In addition, farmers fertilize their huge territories. Once they harvest their crop, the land starts leaking nitrous oxide and methane. No one knows how much greenhouse gases escape from the land into the air, but the amounts could be substantial. Planting cover crops slows down those emissions.[35]

The deadly global warming crisis is anthropogenic in my opinion. That is, I believe that the planet is warming up from additional solar heat caught by greenhouse gases produced by the burning of fossil fuels. As I already said, industrialized farming has a link to the burning of fossil fuels. Corporations modernized the Roman plantation to take advantage of how the world works. They know the world is pursuing agricultural industrialization for reasons of state, corporate power and profit. This policy receives its fuel from the benefits of man controlling nature.

Animal farms are a step in that direction. Cows behind bars are forced into a rigid regimen of feeding, artificial insemination, milking and, in two years, slaughter for their transformation into hamburgers. Perhaps this awful fate has become encoded in the cows' genes. During my visit of the small dairy in the Central Valley of California, I noticed there was a very small green field for the cows to eat grass. Next to the enclosed space holding the cows, there are small mountains of feed. Every so often, a truck would move slowly next to the fence enclosure and spill feed all along the length of the fence. The cows, taking their time, would perpetually be munching on the palletized food. Unwillingly, they had been made into milk and meat machines. The owner said to me he had the dairy since 1969 (more on this in Chapter 5). I told my host about Goldschmidt's findings: large farmers sucking the life out of small towns. They buy their feed, tools and machinery from big manufacturers; they borrow money from large banks; and there are few of them in control of most of the available land. This trade all but kills small business and local taxes; thus, the

---

[34]Pat Costner and Joe Thornton, *We All Live Downstream: The Mississippi River and the National Toxics Crisis* (Greenpeace, December 1989).

[35]Marc Fawcett-Atkinson, These simple farming techniques [like planting cover crops] can curb greenhouse gas emissions, *National Observer*, November 14, 2020.

town eventually shuts down. My friend, however, rejected that argument, telling me he has worked hard for many years, benefiting his hometown. He also admitted he joined the Tulare County Farm Bureau just recently because, for several years, he did not agree with the politics of the Farm Bureau. Thus, this large farmer is his own man.

Arvin and Dinuba in 2012: The Ruin of Rural America

I visited Arvin and Dinuba in the summer of 2012. The downtown of Arvin was desolate. Most of its houses were tiny and temporary. Approaching it was like going through a waste dump. The stench was unbearable. Perhaps the odors came from the dairies on its borders. Most dairies around Arvin and Dinuba are large farms. In the 1940s, Walter Goldschmidt demonstrated that large farms brought ruin to Arvin.

I was taking pictures of what looked like an immense grapevine, when a shining white truck stopped next to me. The driver was a Mexican man working for one of the vast table grape farms of Arvin. He said he had been working for 40 years and now he was the manager of the workers. However, he owned no land, and his children had no interest in following on his footsteps. He saw nothing wrong in few farmers owning vast tracks of land. He disagreed that Arvin was a colony of large farms. Table grape farms and almond plantations, thousands of acres in size, dominate the lonely flat land. The almond trees, in their white flowers, were beautiful. But I would not like being a honeybee pollinating almond trees during the spray season. Farmers use neurotoxic pesticides like neonicotinoids on the trees. Those sprayed trees would be loaded by asphyxiating gases. Behind each blossom there would be death in the form of poison mist or fatal danger in microscopic bubbles full of nerve poison.[36] These pesticides are connected to the chemistry of the nerve agents Europeans blasted each other during World War I. A high percentage of honeybees visiting an almond plantation die. Armies of Hispanic farm workers harvest the crop, moving on trucks or stooped in the fields. They keep this thriving empire of one or two crops alive.

---

[36]Evaggelos Vallianatos, Honeybees and America in Trouble, *Counterpunch*, May 19, 2023.

If you make the effort to find the "town" of Arvin, you see a settlement, not a town. Arvin has no hospital or entertainment. Its public library is open 3 days a week. In fact, even that little library belongs to the county, not Arvin. Its 20,000 people are mostly from Mexico and Central America. They barely speak or understand English. They buy their food from only one Mexican supermarket. Dinuba also has about 20,000 people. Dinuba fares better than Arvin because it has had a stronger agrarian base than Arvin. Dinuba in the 1940s had plenty of small family farms that took care of their community. But in the 72 years since the 1940 study, that base has eroded to the point that in 2012, in the place of small farmers producing good food and wealth for themselves and the community, giant corporate citrus and dairy farms dominate both society and economy. Add to that Wal-Mart, the giant store putting small stores out of business, and the effect is crippling on Dinuba.

I spoke to Joanne Cedbetter, a Dinuba rancher, teacher for 35 years, and a volunteer at the Dinuba Historical Society. This Society is housed in a handsome house that used to serve the Southern Pacific Railroad. Dinuba and other small towns came into being in the late nineteenth century to serve the railroad. A huge room of the "Depot Museum" is pasted with pictures outlining the agricultural history of Dinuba. A large picture highlighted the "raisin day" of 1915. One sees uniformed musicians next to horse riders ready for a parade. Raisins remained at the top of Dinuba crops down to 1960s where, in another picture, Dinuba is called "Raisinland – USA: Home of the Emperor." Joanne Cedbetter told me that Dinuba no longer has a "downtown" or a small business class. She recounted the good old times when, indeed, Dinuba was a thriving rural town with a middle class. She, too, blames the industrialized large farmers for the decline and fall of rural society in the Central Valley. The population of Dinuba, like the population of Arvin, is overwhelmingly Hispanic. This population reversal from white to Hispanics in both Arvin and Dinuba is as ominous as the disappearance of the middle class. The people I met told me that whites were in the vast majority down to the 1960s. Indeed, whites from the dust bowl states of Oklahoma and Kansas would also harvest the crops in the Central Valley. But the takeover of farming in the Central Valley by large farms undid society so thoroughly that now the Central Valley has become entirely a plantation with a tiny number of white farmers owning large portions of land and a vast number of Hispanics doing the work. This class division (of a few masters and many hired workers) is an unsustainable and dangerous precedent in a society

Fig. 3.   Farm workers harvesting and packing vegetables. Coachella Valley, California. Photo by Evaggelos Vallianatos, Slave-Labor Conditions Persists in Coachella Valley, Truthout, April 8, 2011.

still calling itself democratic in my opinion. You cannot have democracy when so few control the land and have to import armies of foreigners to do the work. Farm work is difficult and dangerous. Farm workers often harvest crops that have been sprayed. And other times, helicopters or small aircraft spray biocides over them.[37]

We already said that farm workers are exposed to a variety of hazardous chemicals with potential for harm; their side effects include cancer, birth defects, sterility, behavioral disorders, learning disabilities, brain damage and psychological and neurological diseases.[38] These facts bother me. I do not think we should harm the health of people. I love democracy and I will continue to highlight its advantages like making possible the flourishing of small self-reliant family farmers. These traditional farmers raise wholesome food as much as healthy communities and civilizations, including protecting the natural world. However, this extraordinary fact of the wholesale destruction of rural America by large farmers relying on imported cheap laborers is rarely being discussed because the

---

[37]Krebs, *The Corporate Reapers*, pp. 26–33; Vallianatos, *This Land Is Their Land*, pp. 50–67.

[38]Janette D. Sherman, *Life's Delicate Balance: Causes and Prevention of Breast Cancer* (Taylor and Francis, 2000).

politically-correct sociologists and other academic experts avoid it. After all, what would the alternative be? Redistribute the land to thousands to recreate a small class of family farmers? I would say, yes. And since large farmers still own vast parts of rural America, who would harvest the country's food but the nameless thousands of farm workers from Mexico?[39]

The Grapes of Wrath

On February 11, 2011, I joined a friend professor at Pomona College, Sheila Pinkel, and visited Coachella Valley in California. Once at Coachella, we went to the offices of the California Rural Legal Assistance (CRLA), a federally-funded program designed to do some good to a population at the bottom of American society. Rural Legal Assistance in California and other states remains the last resort of legal advice and representation to extremely poor and disenfranchised rural people. Our CRLA speaker was Megan Beaman, a young woman from Iowa who became a lawyer in order to help farm workers. Beaman started her presentation with numbers. She said 14,000 to 28,000 farm workers moved through the farms of Coachella every year. 57% of those workers were undocumented; 90% of the workers were from Mexico, and some from El Salvador. Beaman then turned her attention to the 3000 to 5000 of farm workers who live year-round in Coachella. These workers, known as Purhepechas, are from Michoacan, a state in Mexico. They speak neither Spanish nor English but their own language. They rent houses that compare with slums in Latin America. Their "trailer park" is on land of the Torres Martinez Desert Cahuilla Indian Reservation. An Indian, Harvey Duro, is the owner of the trailer park also known as Duroville. Duro turned his 40-acre "park" into a cash cow, providing the Purhepechas with the rudiments of housing, with problematic to intolerable conditions for water, garbage disposal, electricity and sewerage.

CRLA came to the assistance of the Mexican farm workers stuck on Duroville and, in 2008, a district court put the town of Harvey Duro at receivership. This means a not-for-profit organization is now running Duroville. This is then one of the main roles of the CRLA lawyer, Megan Beaman. She convinced the court not to demolish the farm worker slum.

---

[39] Vallianatos, *This Land is Their Land*, 137–196.

"Just imagine," she said, "having another exodus of thousands of home-less farm workers." Thus, keeping Duroville intact served a purpose. The farm workers have a roof over their heads and agribusiness has a steady supply of cheap labor and none of the social stigma associated with exploitation. I was astonished that the young and innocent looking Beaman said practically nothing about pesticides, the most persistent, deleterious threat farm workers have been facing in their daily work for decades. I asked her to explain. She said she had no data about pesticide effects and neither did she know of any study about pesticides and farm workers in the Coachella Valley. I reminded her of my experience at EPA and she noted that, yes, pesticides were everywhere in the farms of the Coachella Valley.

To my dismay and palpable anger, I realized how deep the American malaise is about their "invisible" farm workers. First, they refuse to believe that farm workers are harvesting their food. Second, unless they go out of their way – exactly like Sheila Pinkel and her students did – they are not about to see these dark-skinned foreigners picking their food. And third, the few experts who make a living around the broken lives of farm workers are not entirely honest about what they know. Their salaries come from the government or agribusiness that, historically, aim to control rather than care for farm laborers.

I asked the young security officer of the NGO managing Duroville to make sense of the crisis around him. He had just given us an overview of the Mexican community living in Duroville. He spoke about petty crime, beatings of the women by their men, very little interest in education, and the harsh conditions of dire poverty all around him. He said nearly nothing had changed in agriculture since 1939 when John Steinbeck wrote *The Grapes of Wrath*. Steinbeck's novel was the daily reality for farm workers who harvested the wrath of agribusiness. Later in this book, I will discuss that even during the 2020 pandemic, nothing changed to protect farm workers from the additional dangers of the coronavirus as well as the heat and fire waves that crippled the West Coast. The Grapes of Wrath remain intact.

The devastating fires in California in August and September 2020 did not affect the work of farm workers, most of whom were undocumented. The risks of the pandemic and the climate-fueled fires kept enveloping the farm workers but brought no relief to what they did. They knew, and many said it aloud: they couldn't afford to stay home – if they had a shanty they called home. And neither did they dare protest because of the real fear of being fired. Despite the courageous admission of the governor of California, Gavin Newsom, of the climate chaos and emergency hurting and burning the entire West Coast, the State of California, most of the time, allows farmers to continue spraying their crops with neurotoxic and carcinogenic pesticides, no matter the dangers to the health and wellbeing of farm workers who do the harvesting of those crops. Estella Cisneros, director of CRLA, said, "Farm workers have continued to work during this whole time, despite fears of contracting COVID-19 in the workplace, despite fears of getting heat stress while they were at work, and now, despite fears of the dangers that wildfire smoke brought."[40] This is routine in the Sonoma County grape farms of California. Local officials rarely refuse to please the wine owners. The grapes must be harvested at any cost. "We work when there are rains, we work when there is fire, we work in whatever conditions. It isn't the most viable, but it is a necessity to provide for our families," said Gervacio Peña Lopez, a board member of the local Indigenous workers' group Movimiento Cultural de la Unión Indígena (Cultural Movement of the Indigenous Union). Lopez is Mixtec and did farm work for years. "There is no resource we can count on, so there's nothing left but to work," he said.

Lopez is not exaggerating. California officials work with the wine owners as well as the agricultural commissioners of the Sonoma County Farm Bureau and Sonoma County Winegrowers. They repeatedly allowed grape farmers to harvest even when the fires were next door, right in the wildfire-evacuation zones.[41] Agribusiness operates that way. It can appeal to its subsidiary professors at 65 land grant (agricultural) universities to testify that agribusiness means lots of food for America and the world. These professors also turn a blind eye to pesticides and equate them to science.[42]

Some 70 years after Steinbeck's *The Grapes of Wrath*, I found myself in the Coachella Valley and felt like I had traveled back in time. The farm workers harvesting broccoli and a truck moving very slowly formed a human broccoli machine. They worked fast, each worker doing but one thing, and passing the broccoli to the next worker until the vegetable ended up in a box on the truck. The workers looked the same, dressed in

---

[40]*Democracy Now*, September 14, 2020.

[41]Allen Brown, In California's Wine Country, Undocumented Grape Pickers Forced to Work in Fire Evacuation Zones, *The Intercept*, September 6, 2020.

[42]Bosch, *The Pesticide Conspiracy*, 119–128.

the same colors, were all Mexicans, all standing very closely to each other like honeybees. Some gave a glimpse to the students photographing them but, in general, they ignored everything but their mechanical labor. The owner of the broccoli farm, some 35 acres in size, was a young man named Steve Powell, executive vice president of Peter Rabbit Farms. I asked him if he ever thought of converting his agribusiness to organic farming practices. He looked at me and casually said I was living in the past. "We can't do that," he said. I pointed out that being hooked on pesticides was hazardous to those eating his crops, to farm workers, and to himself. He smiled and said the "chemistry" he used was quite specific, targeting destructive insects, his poisons were harmless to people. And with that, he went to the broccoli machine, grabbed some broccoli and handed them to those snapping pictures.

Sheila Pinkel asked Powell pointed but gentle questions about his workers. "What workers," he asked. "The workers you see don't work for me. I have nothing to do with them. They work for a contractor who harvests my crop. I deal with him, not the workers. Years back, we used to hire about 500 workers but no longer. Now the contractors are our lifelines in harvesting our crops."

The Secret Political Life of Lake Apopka

In April 17–18, 2015, I attended a conference in Orlando, Florida. Beyond Pesticides, a non-profit environmental organization in Washington, DC, founded in 1981, sponsored the forum. Before the conference started, some fifty people took a bus tour of rural Florida, especially exploring the lush agricultural region around Lake Apopka, Florida's fourth largest lake. I entered the bus with apprehension. I had heard bad things were taking place in central Florida. Our guide, Jeannie Economos, had been working for the Farmworker Association of Florida for about twenty years. She appeared to me as the embodiment of a person living the ecocide and brutality of the agricultural economy of Florida. She fiercely defends the dignity and human rights of farm workers. Economos knows the history and politics that shape Lake Apopka. She shines in virtue. When listening to her, one has little doubt that behind the facts, there's injustice hovering over all of Florida.

That injustice has two faces. First, Economos said, Florida harmed Lake Apopka in ways hard to fathom. During World War II, Economos said, the state allowed farmers to slice about 20,000 acres of land from the lake. The farmers then converted the rich muck soil into vegetable factories. But the pollution of those industrialized vegetable farms nearly killed the lake. So, in the 1980s, the state of Florida bought off those lake farmers, hoping without hope to save the lake. Lake Apopka in 2015 was moribund. Its wildlife has been dwindling, verging on extinction. The farmers' addiction to biocides and synthetic fertilizers, Economos explained, nearly wiped out lake fish and birds alike. In 1998, about 1000 fish-eating birds died from poisoning. The poisons that killed the birds were primarily DDT-like sprays.

The alligators of Lake Apopka mirror the deleterious state of the lake. The maladies of alligators are maladies of ecosystems poisoned by industrial agriculture and other industries surrounding the lake. The alligators failed to reproduce. When the conference began, Professor Louis Guillette, who taught for many years at the University of Florida and, in 2015, was teaching at the Medical University of South Carolina, explained how and why poisons in the lake cause reproductive and sexual abnormalities to alligators and other water animals. Pesticides castrate alligators. Guillette should know. He made his scientific career by studying the decline of alligators of Lake Apopka. Another scientist, Professor Tyrone Hayes of the University of California-Berkeley, told us that pesticides cause havoc to wildlife, especially amphibians. Hayes explained in graphic detail how the "popular" weed killer atrazine disrupts the lives of frogs. Atrazine in minute amounts feminizes male frogs, reducing their potential to procreate. Hayes said those who spray atrazine have thousands of times more atrazine in their body than the amount required to reduce the reproductive ability of frogs. The atrazine in humans comes from drinking water and food, according to Professor Hayes. He also did not mince words telling us how the owners of atrazine have been harassing and persecuting him.

Out in the fields near Lake Apopka, we could see land littered with medical incinerators, tree, vegetable and flower "nurseries," factories, large farms, and real estate development. I was told that these businesses, especially the abandoned factories that used to manufacture pesticides, have a dominant and toxic footprint on the natural world. The chemical factories and especially the manufacture and use of organchlorine pesticides contaminated the sediment and watershed of Lake Apopka. In order to remedy the situation, the US Environmental Protection Agency called Lake Apopka a "superfund" site.[43] It's not easy to reverse the ecological conditions at Superfund sites. Spending billions and decades, the government out-sources the cleaning up of the worst pollution. No wonder our tour of Lake Apopka was coined a "toxic" tour.

The Oppressive Life of Farm Workers

The second fact about present-day Lake Apopka is the lives of the farm workers; the thousands of them living around the lake. I was again told during the tour, that these farm workers live lives filled with poverty, disease and death. Economos spoke with anger about the oppressive lives of farm workers. These destitute people, she said, are largely black and Hispanic. They have been working for decades, making wealth for Florida's rural oligarchy. Economos quoted a slogan that a rural oligarch was supposed to have said: "We used to buy slaves; now we rent them." Nevertheless, employing or renting "slaves-like workers" did not save the Apopka farmers. The state put them out of business in 1998. However, the state did nothing for the farm workers who had worked in the farms of Lake Apopka for about fifty years. Suddenly, farm workers lost everything. Some of them even became homeless. "The state spent hundreds of millions on the lake," Economos said, "but not a dime on the farm workers." And the University of Florida, she accused, is not better than the state. It has a Food Research Institute next to the lake, but it does not see the farm workers. It can point to the EPA issuing its Farm Worker Protection Standards, a lipstick measure leaving the toxic environment of industrial agriculture intact.

American Feudalism

The plight of the farm workers has spread to rural towns encircled by the enclosures of giant farms. When a rural town is in the midst of large farms, Goldschmidt said (and cited above), it becomes like a transient camp, its public services all but disappear, and small businesses shut down.

---

[43] The US Environmental Protection Agency defines extremely polluted sites as Superfund sites. EPA has designated Lake Apopka as a Superfund Site because of "elevated levels of some organochlorine pesticides" in its sediments and part of the lake's watershed: US EPA, Health and Environmental Research Online, 2016.

Democratic life declines. The only jobs left in the town are jobs serving agribusiness. Goldschmidt described these symptoms of disease now afflicting Arvin and Dinuba – Arvin being in a state of stupor, Dinuba barely hanging on. As for farmers with farms as large as 160 acres or more, whenever they find themselves in the company of agribusiness, they are forced to abandon farming. That also happened to the small farmers of Arvin and Dinuba.

In 1983, another researcher, Dean MacCannell, professor of rural sociology at the University of California-Davis, issued a severe warning that complemented the warning of Goldschmidt: the size of farms, MacCannell said, matters in agriculture. Large farms destroy rural America. He said agribusiness policies "cut against the grain of traditional American values." His studies showed that giant farmers were becoming America's "neo-feudal" lords who, with government assistance, were converting rural America into a Third World of poverty, injustice, exploitation and oppression. When large farms are in or near small farm communities, he says, they ruin the rural communities, sucking all life out of them: "In the place of towns which could accurately be characterized as providing their residents with [a] clean and healthy environment, a great deal of social equality and local autonomy," he explains, "we find agricultural pollution, labor practices that lead to increasing social inequality, restricted opportunity to obtain land and start new enterprise[s], and the suppression of the development of [the] local middle class and the business and services demanded by such a class."[44]

A few years later, in 1990, Linda Lobao of Ohio State University published the results of her sociological study on the effects of industrialized farming on rural communities. She picked up where Goldschmidt had left off. Her data came from 3000 US counties. Like Goldschmidt, she found decay in communities dependent on large farms. In 2006, Curtis Stofferahn of the University of North Dakota updated the work of Lobao. In summarizing the findings from 50 years of social science research, he reached the following conclusions: "[Industrialized agriculture] disrupts the social fabric of communities... poses environmental threats where livestock production is concentrated; and is likely to create a new pattern

---

[44]Dean MacCannell, Agribusiness and the Small Community (background paper to *"Technology, Public Policy and the Changing Structure of American Agriculture,"* US Office of Technology Assessment, US Congress, 1983).

of 'haves and have nots."[45] In other words, Lobao and Stofferahn confirmed the conclusions of Goldschmidt; the fear of the 1884 California
State Agricultural Society: that large farms are bad for society and
democracy.

In the first chapter I spoke about Greek small family farmers responsible for the foundations of the polis (city-state). I connected the farmers
to the political institutions of law and liberty that blossomed into the religion, art, philosophy, science, literature, and social life that made Greek
civilization so powerful and lasting. The Roman conquest of Greece in
146 BCE undermined small Greek family farmers. Large farms became
the icons of the new imperial state in both Greece and the Roman Empire.
Victor Davis Hanson, a family farmer and former professor of Greek at
California State University, Fresno, studied the political significance of
Greek farming and he is right to argue that agrarianism made Greek political culture. "Only a settled countryside of numerous small farmers," he
says, "could provide the prerequisite mass for constitutional government
and egalitarian solidarity." [46] The lesson of Greece and Rome suggest
that the takeover of farming by large interests is not entirely a function
of the industrial revolution. Both Hanson and Goldschmidt – steeped in
the theory and practice of the democratic harvest of family farming and
the authoritarian and other deleterious consequences of large-scale corporate agriculture – are bitter about the taking over of the countryside of
America by the very form of factory farming they dislike. Hanson criticized giant agriculture in a personal account[47] of his own efforts to survive
as a family farmer. In it, he talks as if he is writing a "postmortem" to the
millennium "agrarian Armageddon" responsible for the obliteration of
family farming in the United States.

I don't think Armageddon – agrarian or otherwise – has a place in the
fate of agriculture. But I understand the fury that has consumed so many
family farmers[48] and others who believe passionately in the ability and

---

[45] Curtis W. Stofferahn, *Industrialized Farming and its Relationship to Community Well-being: An Update of a 2000 Report by Linda Lobao* (State of North Dakota, Office of the Attorney General, September 2006) pp. 6–11, 30–32.

[46] Victor Davis Hanson, *The Other Greeks: The Family Farm and the Agrarian Roots of Western Civilization* (New York: The Free Press, 1995) p. 27.

[47] V. D. Hanson, *Fields Without Dreams: Defending the Agrarian Idea* (New York: The Free Press, 1996), p. xi.

[48] G. Logsdon, *At Nature's Pace: Farming and the American Dream* (New York: Pantheon Books, 1994), p. xi.

clear necessity of eating our food from the family farmer's field, and, just as importantly, hold true that several million family farmers throughout the land in America are a must for the preservation of freedom and a democratic form of government.[49] The failure of that democratic and agrarian dream is causing anxiety to those who, like Hanson and Goldschmidt, are certain of the rigid alternatives to family farming.

Goldschmidt testified before Senator Gaylord Nelson on March 1, 1972 – twenty-six years after another Senator, James E. Murray, made it possible that his path-breaking study of the destructive effects of large farms on family farming and human communities saw the light of the day. Goldschmidt said Congress ought to know more about "the increased encroachment of agribusiness on American rural life" primarily because he was convinced that "corporate farming creates an urbanized and impoverished rural community." He also accused agribusiness of decimating the number of family farmers and the federal government for making all that possible – particularly with its policies of agricultural support and farm labor.[50]

I met Goldschmidt in 1987 in Florida at an academic conference. I asked him if the country had a chance to break up agribusiness and distribute its lands to small family farmers; in other words: do to America's large farmers what the American general, Douglas MacArthur, did to the large farmers of defeated Japan. He said my proposal was not feasible without a willing dictator-president. I was stunned and rejected his argument. But, deep inside, I knew he probably was right. I was in Washington, DC, in the mid-1970s when the Small Business Committee of the Senate held extensive hearings on the fate of the family farmer. Senator Gaylord Nelson of Wisconsin was one of the very few voices behind Congress' belated interest on family agriculture. Witness after witness would describe the infractions of agribusiness against family farmers

---

[49] Krebs, *The Corporate Reapers*, pp. 23–33; Bob Bergland, *A Time to Choose: Summary Report on the Structure of Agriculture*, pp. 1–12; Vallianatos, *This Land is Their Land*, pp. 262–286.

[50] W. Goldschmidt, Research into the effects of corporate farming on the quality of rural community life, *Role of Giant Corporations*, Hearings Before the Subcommittee on Monopoly of the Select Committee on Small Business, United States Senate, Ninety-Second Congress, First and Second Sessions, Part 3, Corporate Secrecy: Agribusiness, November 23 and December 1, 1971; March 1 and 2, 1972 (Washington, DC: US Government Printing Office, 1973), pp. 3925–3947.

and rural America. Yet, listening to the testimony of the defenders of family agriculture, I felt like I was at a funeral. It made no difference how passionate the testimony was. The victim was already dead. And no one was there to listen. Congress and USDA pretended the tragedy of family farming did not exist. In 1992 we heard that "Abandoned farmsteads and closed businesses haunt rural Kansas."[51] Marty Strange, an agricultural policy analyst for at least 20 years with Nebraska's Center for Rural Affairs, said that a fundamental demographic shift was taking place in the farm communities of America: Farmers were getting few and old, fast. Strange saw nothing good from this unsettling of the land. He said that while "wholesales of fertilizer" increased during the 1970s and 1980s, "sales of groceries, shoes, haircuts and everything else that has to do with people fell. This has left rural main street with little more than dealers, brokers, franchisers and agents – businesses designed to siphon money of the community."[52]

Giant agriculture is an industrial system that has very little to do with traditional farming. It relies on "producing" one crop at a time in expanses of land – e.g., plantations and huge spreads – feeding that one crop with synthetic fertilizers, and protecting it from insects and diseases with powerful toxins. Giant agriculture is also concentrated livestock operations, and food processing and marketing of food. Moreover, industrialized agriculture mines the land. Nearly half of the world's wrecked agricultural soil, some of it moderately eroded, some of it severely degraded, is in Africa.[53] Yet economic analysis, it seems to me, obscures the degradation and, sometimes, destruction of the natural world in the absence of which no agriculture is possible.

Loss of Crop Genetic Diversity

The persistent policies of conventional agriculture of using only a handful of crops to produce most of the world's food – in largely huge farms displacing the tiny but immensely rich in biological and cultural diversity

---

[51] Jerry Jost, Looking to the future, *Sustainable Farming News*. February 1992 (Kansas Rural Center, Whiting, Kansas).

[52] Marty Strange, An open letter to the sustainable agriculture movement, *Center for Rural Affairs Newsletter*, December 1990 (Walthill, Nebraska).

[53] United Nations Environment Programme, *Global Environment Outlook* (New York: Oxford University Press, 1997), p. 26.

family farms – are responsible for the tragic loss of a considerable amount of genetic resources for food and agriculture. In 1903, there were 13 known varieties of asparagus. By 1983, only one variety of asparagus existed. Nearly 98% of asparagus varieties became extinct between 1903 and 1983. There were similar losses for carrots, radishes, and lettuces. In 1903, there were 287 varieties of carrots, more than 460 varieties of radishes, and about 500 varieties of lettuces. By 1983 there were 21 varieties of carrots, 27 varieties of radishes, and 36 varieties of lettuces. Clearly the twentieth century was a terrible century. Many varieties of crops did not survive the cash croppers and scientists who were set on influencing nature. From 1903 to 1983, *about 75% of vegetables and other varieties of food crops disappeared.*[54] According to Hugh Iltis, the world-renown botanist at the University of Wisconsin, cash crop agriculture causes biological genocide and utter devastation in the tropics.[55]

Cash cropping as well as the spread and globalization of Western culture had another dramatic effect on human culture. They killed cultural diversity, particularly the disappearance of spoken languages among indigenous peoples and small ethnic groups. The United Nations Environment Program estimates that *the globalization of Western culture is likely to kill about 90% of the world's 5,000 to 7,000 languages by the end of the 20th century.*[56] The situation is so bad in the impoverishment of both cultural and biological diversity that determine what people have been worshipping, growing, and eating for millennia that one can describe the loss of agricultural biodiversity and the erosion of cultural diversity as a biological and cultural meltdown.[57]

---

[54] Cary Fowler and Pat Mooney, *Shattering: Food, Politics, and the Loss of Genetic Diversity* (Tuscon: University of Arizona Press, 1990), pp. 64–67.

[55] Hugh Iltis, Extinction is Forever, *Resurgence*, November/December 1997, pp. 18–22; Fowler and Mooney, *Shattering*, pp. ix–xiv, 83.

[56] Luisa Maffi, Linguistic Diversity, in Darrell Addison Posey, ed., *Cultural and Spiritual Values of Biodiversity: A Complementary Contribution to the Global Biodiversity Assessment* (London: Intermediate Technology Publications on behalf of the UN Environment Programme, 1999), pp. 21–35.

[57] Hope Shand, *Human Nature: Agricultural Biodiversity and Farm-Based Food Security* (Ottawa, Canada: Rural Advancement Foundation International, December 1997), pp. 1–9. This report was prepared for the UN Food and Agriculture Organization.

Why Small Traditional Farms Are Beautiful

In contrast to the bad effects of large farms, small family farms draw from eons-tested wisdom of how to grow food. Nature is their model. Their seeds are full of biodiversity, growing into nutritious food. They often produce per acre equal or more food than large-scale farms.[58] They cultivate merely a quarter of all farmlands.[59] Yet, according to a small international non-profit agrarian organization, Grain, *small-family farmers are basically feeding the world.*[60] They grow a variety of crops, which incorporate the power of nature in minimizing harm from insects. Ecologists study small traditional farms. In fact, a new field of science, agroecology,[61] is enriching traditional small-scale farming with modern ecological science.

The Power of Agroecology

Miguel Altieri, emeritus professor of agroecology at Berkeley, has been criticizing giant agriculture for decades. I met him in the mid-1990s when the EPA seconded me to the United Nations Development Program in New York. Altieri is a rare academic scientist who did not mind working with his hands and learning from small farmers. He put into practice, the latest findings of ecology with the insights and wisdom of traditional farmers and formed the science of agroecology. He has no doubt that, in an age of pandemics, agroecology is the answer. At a time when the world is threatened by climate change and the pandemic, Altieri says, agroecology has the potential to guide and motivate small farmers to produce most of the food we need, produce it locally, and revitalize our communities and the health of the natural world. For this to happen, we must spread from farmer to farmer "pedagogic strategies" and create "agroecological lighthouses" for the revival of traditional knowledge and methods of growing food. This process will inspire us to think of eating as an ecological and

---

[58] Vallianatos, *This Land is Their Land*, pp. 262–286.

[59] Peter M. Rosset, The multiple functions and benefits of small farm agriculture, *Food First*, September 1999.

[60] Grain, Hungry for land: small farmers feed the world with less than a quarter of all farmland, May 28, 2014.

[61] Miguel A. Altieri, Agroecology, Small Farms, and Food Sovereignty, *Monthly Review*, July 1, 2009.

political act. "COVID-19," he says, "is reminding us that disrespectful treatment of plant and animal biodiversity has consequences."[62]

My hope is that Congressional Democrats will listen to Altieri and the attractive options of agroecology for the peaceful transformation of America's agriculture and food, making them healthy to us and the natural world. The 47th President should consider supporting small family farms, which can start growing a large variety of healthy food. President Biden made a mistake in bringing back to power the Obama administration's secretary of agriculture, Tom Vilsack. He is a former governor of Iowa, a dairy lobbyist, and a corporate man who was unsuitable for this critical time of transition away from the adverse effects and consequences of petroleum farming and animal farms of giant agriculture. His leading position encouraged and supported a ceaseless growth of the power Big Ag has over family farmers and rural America. The Obama-era scandals are symptoms of where Vilsack stands. From 2009 to 2017, several cases of corruption hit the headlines but without impact on the policies of Vilsack. These "scandals erupted in five of the biggest checkoff programs: beef, pork, dairy, egg, and soybean. Hundreds of thousands of misappropriated dollars were involved, and so were accusations of racketeering, illegal lobbying, congressional inquiries, multiple lawsuits, even threats of physical violence."[63]

The checkoff programs are USDA programs taxing farmers and using the money to promote agricultural products at home and abroad. The practices under his watch weakened efforts to fight and limit the power of Big Ag, as suggested by one of his own senior executives, J. Dudley Butler. In a December 22, 2014 letter he sent to Vilsack, he said: "Many producers, conservative and progressive, believed your promises and were hopeful for a new day at USDA. Some took brave stances based on your promises to their own peril. Instead, they got more of the same – an agency controlled by the big food companies and the big meat packers as well as their minions... Your lack of leadership has ensured that independent cattle producers will continue to be systematically pushed toward the slaughterhouse of vertical integration."[64]

---

[62] Miguel A. Altieri and Clara Ines Nicholls, Agroecology in Times of COVID-19, 2020.

[63] Clint Rainey, Vilsack's checkoff problem, *Food and Environment Network*, April 22, 2021.

[64] Former GIPSA [Grain Inspection Packers and Stockyard Administration] Administrator Calls for Secretary Vilsack's Resignation, Organization for Competitive Markets, December 22, 2014.

Fig. 4.   Strip of buckwheat flowers in a vegetable farm. They provide pollen and nectar to the natural insect enemies of pests. Courtesy of M. A. Altieri, Impact of flowering backwheat on Lepidoptera cabbage pests, *Biological Control*, Sept. 2005, 34 (3) 290–301.

President Biden's choice of Vilsack reveals more about Biden. You would think that the climate emergency would bring Vilsack and Biden to their senses. On the contrary, Vilsack and his boss, Biden, supported agribusiness policies that mirrored petrochemical priorities. More on this below on the chapter on climate change. Suffice it to say that the American Meat Institute, the Dairy Export Council, and giant pesticide and seed companies like Syngenta, Bayer, and Corteva joined Vilsack's machine.[65]

Nevertheless, agroecology is a path toward for a democratic and healthy transition of growing food in America, and a scientific approach to fighting pandemics.[66]

---

[65]Tom Philpott, Agricultural emissions are not on the menu at COP 26, *Bulletin of the Atomic Scientists*, November 5, 2021.
[66]Miguel A. Altieri, The scaling up of agroecology: spreading the hope for food sovereignty and resilience. *SOCLA*, May 2012.

Agroecology encourages the small family farmer to attract to their land insects that feed on crop-harming insects (Fig. 5).

Small traditional farmers (organic small family farmers in America and Europe and farmers in the tropics) need more land. Their farms have none of the technology and power baggage of large, industrialized farms.

Fig. 5. A syrphid fly, predator of aphids; the latter are pests to crops. Courtesy of Miguel Altieri.

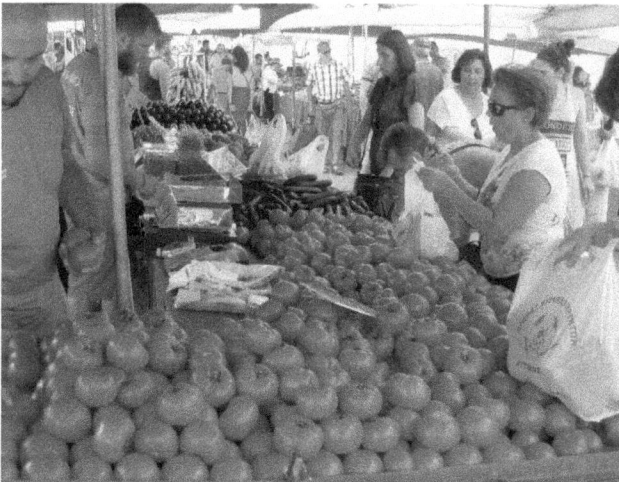

Fig. 6. Organic tomatoes from Crete in a farmers' market in Vrilisia, a suburb of Athens, Greece, June 7, 2018. Photo by E. Vallianatos, *Greece: From Burning to a Green Future*, *Counterpunch*, August 1, 2023.

They produce nutritious food and, as I have argued above, give democracy as well as meaning and vitality to rural communities. They coexist in harmony with the natural world, being a potential prelude to ecological civilization. Their knowledge and example might heal the otherwise harmed territory of factory agriculture. That's why small traditional organic farms are beautiful.

The next chapter examines the origins of pandemics, especially those associated with the destruction of nature, and those emerging from animal farms, all seen in the context of climate change.

4 On the Grip of Disease

Pebble mosaic depicting the Greek hero Bellerophon riding the flying horse Pegasus killing Chimera, c. 300–270 BCE. Archaeological Museum of Rhodes, Greece. Public Domain.

This image is symbolic that Chimera was a monster that had no place in Greek society. Similarly, the lab-engineered chimera is a powerful monster we must slay, too.

Plague and Civilization

The consequences of the COVID-19 pandemic included human isolation, illness and death in 200 countries around the world. It was a pandemic with the potential of vast death and partial destruction of national and world economies. In addition, it increased the profits and power of national and global corporations. Brookings, a non-profit, non-governmental organization in America, reported that "shareholder wealth soared," while "workers were left behind."[1]

A bacterial or viral plague raises its devouring and catastrophic head every so many decades or centuries, especially when humans inevitably disturb the natural world – though not necessarily with the same intensity or virulence. In 2020, the UN warned: "Humanity stands at a crossroads with regard to the legacy it leaves to future generations. Biodiversity is declining at an unprecedented rate, and the pressures driving this decline are intensifying. None of the… Biodiversity Targets will be fully met… The COVID-19 pandemic has further highlighted the importance of the relationship between people and nature, and it reminds us all of the profound consequences to our own well-being and survival that can result from continued biodiversity loss and the degradation of ecosystems."[2]

The historical record of plagues is muddled. Like us, past societies under the existential stress of pandemics, failed to keep records, much less accurate records. In many instances, past and present, rulers, medical bureaucrats, and journalists ignored the reality. Political and economic oligarchs fight for survival and supremacy. The picture that survives death is often incomplete and biased, exactly like the story victors tell after a war.

The Plague among the Greeks

The case of the plague in Greek history may still give us pose for reflection. The Greeks gave diseases precise names. They called plague *loimos* (pestilence)*, nosos* (disease, sorrow, suffering), and *phthora* (destruction, decay, mortality, death). The plague made its first appearance among the Greeks as a weapon of divine wrath. The god Apollo used pestilence to punish the Greeks for offending his priest, Chryses. In the beginning of

---

[1] Molly Kinder, Katie Bach, and Laura Stateler, Profits and the Pandemic, *Brookings*, April 21, 2022.

[2] UN Global Biodiversity Outlook 5, Report 2020.

the first book of the *Iliad* by Homer, Agamemnon, commander-in-chief of the Greek troops in the Trojan War, insulted the priest of Apollo by refusing to give back his daughter, whom he had captured in a raid. The priest knelt in front of Agamemnon and begged him to release his daughter. But Agamemnon told the priest to get out of his sight as quickly as he could, lest he lost his patience. The frightened priest ran away from the Greek camp and went home. He immediately prayed to Apollo to punish the Greeks. He reminded the god he had built a temple to honor and worship him, offering him rich sacrifices. "Make the Greeks pay for my tears," he appealed to Apollo. According to the story, Apollo, the Smithian god of plague, listened to Chryses. The angry god rushed "like night" out of Mt. Olympus in Thessaly and landed in the Greek camp in Troy. He started shooting his invisible plague arrows at mules, dogs and soldiers. The dead fell to the ground for nine days, and fires everywhere burned their bodies.[3]

This plague came to an end only by appeasing Apollo. Achilles "of the swift feet," the greatest hero of the Trojan War, asked Kalchas, the "blameless" seer accompanying the troops, to reveal the cause for Apollo spreading the plague among the Greeks. Kalchas said Apollo was very angry because of the way Agamemnon had treated his priest, Chryses. The Greeks, Kalchas said, should return "the glancing-eyed" daughter of the priest to him and sacrifice 100 cattle to honor Apollo, who would then cease his biological warfare against them.

Achilles and Kalchas prevailed. The Greeks returned the daughter of the priest to her father. Apollo stopped shooting his disease arrows against them.[4]

During the ninth year of the Trojan War, however, the plague returned to the Greek camp and Troy because the pestilence was ravaging the neighboring cities in the Hellespont. At this moment, Palamedes, a genius of craftsmanship, warfare, astronomy, and wisdom, convinced the Greek soldiers to fight the plague with a new diet and vigorous health. He replaced the eating of meat with dried fruits, nuts, and vegetables. He also organized a rowing competition and athletic games outside the narrow and unhealthy camp. He even convinced Agamemnon to reward the victors

---

[3] Homer, Iliad 1.9–10, 26–53, 370–385.
[4] Homer, Iliad 1.285–325.

with prizes. The soldiers avoided for the most part the deadly effects of the plague and honored Palamedes.[5]

The next plague incident among the Greeks was a bacteria pestilence. It happened about 800 years after the Trojan War. This was late fifth century BCE when Athens and Sparta were at each other's throats, fighting the destructive and corrosive Peloponnesian War. In the second year of the war, 430–429 BCE, the plague arrived in Athens. The timing of the disease could not have been worse. Rural Athenians from Attica had abandoned their villages, becoming war refugees living behind city walls. Spartans were burning their homes, crops and trees. The walls of Athens went all the way to the port of Piraeus, Athens' chief port. The deadly epidemic reached Athens all the way from Africa through the port of Piraeus. Thucydides, Athenian historian and author of the masterpiece, *The Peloponnesian War*, survived the disease. His description of the malady (below) is dispassionate and, very possibly, accurate.

Disorder of the Seasons

By late fifth century BCE, Greek medicine was on its way to becoming a healing process based on science. The medical hero and god, Asklepios, dating from the Heroic Age of about 1300 BCE, had built a tradition of divine healing and health protection from the careful observation of nature. His two sons, Machaon and Podalirios, fought in the Trojan War as leaders of Thessalian troops.

Asklepios' greatest pupil, Hippocrates of Kos, flourished in late fifth century BCE. He spoke the language of science and diet. He saw trouble, perhaps the rise of plague, when in hot temperatures the rains came down violently. He advised physicians to study the environmental causes of health and disease: the rise and setting of the sun and constellations, the decisive importance of climate and the seasons, the quality of water, frequency of rains, winds, as well as the fertility and availability of land and food. He said there was no divine disease, only disease caused by bad diet and the alteration of climate and nature. "Natural cause," he said, drives disease. Nothing happens without a natural cause.[6]

---

[5]Philostratos, *Heroikos* 33.14–42. Philostratos was a Greek philosopher who was born in the Aegean island of Lemnos and lived from 170 to 247 in our era.
[6]Hippocrates, *Airs, Waters, Places* Section 22.

The Plague of Athens

However, a plague was and is a dreadful pestilence with no easy recipes for cure. Thucydides knew this, which was why he was so clinical in his details of the symptoms of the disease that killed the Athenian leader, Pericles. Thucydides' comments on the pestilence remind us of our vulnerability and almost accidental survival. Here are the highlights of his description of the plague:

"Not many days after the Spartans and their allies were scorching Attica, the plague struck Athens. People said the disease had already spread widely, including in [the Aegean island of] Lemnos. The plague was of enormous magnitude and caused a great loss of life. Physicians did not have the knowledge for treating the sick and dying. They were no match for it and died in droves. Other efforts of appealing to the priests at the temples and oracles did no good.... The symptoms of the plague included high fever and redness and inflammation of the eyes. The throat and tongue were usually swollen with blood. Breathing became irregular and foul-smelling. Sneezing and hoarseness made everything else worse with extensive coughing. The plague was also responsible for heaving, vomiting, great thirst, insomnia, and restlessness... Helping each other became deadly. Healthy people assisting those in distress were infected and died like cattle... The effects of the plague became even more severe because of the crowding of Athens with Athenians from the villages of Attica. They had no homes but lived in huts, which became stifling in the hot summer. The death rate was high, the dead would fall over other dead in the streets and temples... This plague, and the chaos it brought about, sparked lawlessness in the city. Infected Athenians broke the law and stopped worshipping the gods. Nothing made any difference to those who were dying or on the verge of death."[7]

Almost 400 years later, the Roman poet Lucretius revisited the plague of Athens. Lucretius lived in the first century BCE. His poem on the universe praised science to the heavens. It summarized in Latin verse, the Atomic Theory of fifth century Greek natural philosophers Democritus and Leucippus. They taught that everything in the universe is made up of atoms and empty space. However, Lucretius extracted atomic physics not from Democritus but from Epicurus, a Greek natural philosopher of the

---

[7]Thucydides, *The Peloponnesian War*, 2.47–53

mid-fourth century to early third century BCE. Epicurus blended atomic physics with the virtues of science, study and love of nature, the wise conduct of a prudent life lived in pleasure and doing good.[8] The gods indeed exist, but stay out of human affairs. Epicureanism became popular among the Greeks and Romans.

Why would Lucretius write about the plague of Athens? He was an ambitious man with a good command of Latin and Greek as well as access to Thucydides. His ambition was to explain the nature of the Cosmos, including the nature of disease. So, what better way than paraphrasing Thucydides? He might also have had other sources that complemented Thucydides. We do not know. But we know Lucretius had read Greek literature widely, probably including the works of Hippocrates. Lucretius says there are good and bad atoms. Bad atoms cause disease and death. They do that by upsetting the balance of the atmosphere; the air becomes infected; plague comes in the form of clouds and mists, including heavy rains in hot sunlight. Infected air spreads disorder. Humans then breathe the air from a tainted atmosphere and suck in the plague. This "fatal tide of pestilence" wrecked Athens. The disease, Lucretius says, came from Egypt and dropped on the Athenians who "began to surrender, battalions at a time, to sickness and death." Once a man fell ill, "he lost heart and lay in despair as though under sentence of death. In expectation of death, he gave up his life there and then." The "contagion of the insatiable pestilence... heaped death on death."[9]

By the time of Lucretius, medicine was on its way downward. Galen, a Greek physician of the second century and a polymath genius, put temporary brakes on the decline of medicine. Like Hippocrates, he thought heat and moisture cause chaos in the atmosphere, and result in plagues triggered by the disorder of the seasons, whereby violent storms clash with burning heat.[10]

---

[8] Diogenes Laertius, *Lives of Eminent Philosophers*, 10.39–50, 122–126.

[9] Lucretius, *On the Nature of the Universe*, 6.1089–1239, tr. R. E. Latham (New York: Penguin Books, 1994).

[10] Galen, *Mixtures*, 1.529–532.

A Bad Time for Humanity

After Galen, humans afflicted by plagues dug into religious superstition and suffered massive death rates.[11] In the sixth century, a rat fleeing from Egypt spread a plague to Constantinople, capital of the Eastern Roman Empire (medieval Greece). From Constantinople, the pestilence spread East- and Westward, killing something like 25 to 100 million people in the next 50 years.[12] The 1347–1350 plague was a pandemic of the Black Death. Ships from the Black Sea carried the disease and goods to Messina, Sicily. The pestilence spread to Europe where, in five years, killed a third of the European population. Giovanni Boccaccio, 1313–1375, an ardent Italian Renaissance scholar, was caught in the spreading plague in Florence, a great Italian city state with exceptional traditions of classical learning, the teaching of Greek, banking, and the growing and weaving of wool. The year was 1348 and Florence suffered unspeakable calamities, destruction, dramatic decline and death. Boccaccio, an important Renaissance humanist, lived in Florence and left us a memorable story of the Black Death. He said the plague killed people in the streets of Florence and in their homes. Boccaccio reported in his story, cited below, that neighbors discovered dead people next door from the stench of rotting corpses. Fear of the Black Death was so intense that brother abandoned brother and parents ran away from their children. Society disintegrated. Florence rapidly collapsed. Its political, economic, and religious institutions were wrecked. Even funerals gave way to the dumping of the dead into mass graves. Civilization turned upside down.

The chaos that changed Florence into a living hell gave Boccaccio the opportunity to write his masterpiece, *The Decameron*[13] (Ten Days), a collection of insightful and gripping stories about love, wit, deception, fortune and morality, told through a group of ten people escaping the Black Death from Florence – in other words, a commentary on pestilence and civilization. Boccaccio wrote this book as a frame narrative, where ten young people, seven women and three men, escaping to a villa in

---

[11]Graham Kennedy, *The Global Community Catches a Virus*, New Zealand Center for Global Studies, March 31, 2020.

[12]Ishaan Tharoor, The Plague of Justinian, *Time*, October 26, 2010.

[13]Giovanni Boccaccio, *The Decameron*, tr. Wayne A. Rebhorn (New York: W. W. Norton, 2013).

the countryside to avoid the Black Death, each told a story every day, resulting in a total of 100 tales. Many of these stories revealed the reality of the Black Death and the dreams of a potential future. It also showcased the discrepancy between religious and social ideals of the time with the more pragmatic, human behaviors when faced with a plague.

Boccaccio did not like clerical Christianity and tried to resurrect Hellenic and Roman antiquity and its many gods. He praised the pleasures of the body, not the tortures of monasticism. But the plague threatened everything. The misery it created broke his heart. He started the first story of *The Decameron* with a universal appeal for "compassion for those who suffer." The stories recounted the lost and plague-ravaged civilization of Florence. They offered a vision of a humane, compassionate, and rational world full of desire, magnanimity, intelligence, and good fortune; what the Greeks would call "*kalon k'agathon*," or "the beautiful and the good." Under the shadow of the plague, that world came into being to some degree during the Renaissance of the fifteenth century, during which a great effort was exerted to learn from and be inspired by the Greeks and Romans.

Barbara Tuchman, historian, and author of *A Distant Mirror: The Calamitous 14th Century*, writing under the equally black clouds of the calamitous Cold War of nuclear bombs of the twentieth century, saw the Black Death as "the most lethal disaster in recorded history."[14] She said the first 50 years of the fourteenth century leading to the bubonic plague of 1347–1350, just like the first 50 years of the twentieth century, were "violent, tormented, bewildered, suffering and disintegrating."[15] This was, in fact, the distant mirror of our times. Swap "Black Death" for "Nuclear Bombs," and the similarities are astonishing. The Black Death lasted for three centuries, killing an unknown number of millions of people, ranging from a third to half of the population of Europe.[16] This pestilence culminated in the London plague of 1665.[17] Nevertheless, the Black Death of the fourteenth century did not change the hegemonic and class policies of Europe at home and abroad. Industrialization and the

---

[14] Barbara W. Tuchman, *A Distant Mirror: The Calamitous 14th Century* (New York: Alfred A. Knopf, 1978), pp. xiii.

[15] Tuchman, *A Distant Mirror*, pp. xiii.

[16] See Black Death article in Wikipedia.

[17] The Great Plague of London, 1665, *Contagion: Historical Views of Diseases and Epidemics*, Harvard University Library.

enclosures of land grabbing from European small farm owners, the massacre of the indigenous people of Africa, Asia and the Americas, proceeded at neck-breaking speed. Cities in Europe and in European colonies became fountains of coal-burning mills for child labor and crushing poverty and inequality.[18]

During the middle of the nineteenth century, in 1855, another plague, known as the Third Pandemic hit the world. This pandemic started in Yunnan, China. The impact was long-lasting and the disease spread to Guangzhou and Hong Kong. During the pandemic in Hong Kong, in 1894, Alexander Yersin discovered the bacterium, *Yersinia pestis,* which was probably the only good thing that happened.[19] Due to the Hong Kong's role as a global port, that pathogen reached the rest of the world, killing about 12 million people.

The WWI influenza (flu) pandemic, better known as the Spanish flu, probably infected "about 500 million people or one-third of the world's population." The plague probably killed "50 million worldwide with about 675,000 occurring in the United States."[20] It lasted from 1918 to 1920. It was possible that the vast scale of the war resulted in the vast consequences of the flu-pestilence.

Bush Meat

The late twentieth century and early twenty-first century gave birth to viral pandemics of the Ebola and SARS (Severe Acute Respiratory Syndrome) variety. These were plagues transmitted by fruit-eating tree bats and cave horseshoe bats. These bats did not seek humans, but humans sought to eat them. I remember visiting Nigeria in the late 1990s as an observer to an election. We spent time in Abuja, capital of the country built at the center of Nigeria. We stayed at the Abuja-Hilton where, the daily diet included, among numerous luxurious foods, "bush meats." I asked the chef what kind of food was hiding in the bush meat, and he only smiled. I did not touch bush meat or meat from domesticated animals.

---

[18]Alfred W. Crosby, *Ecological Imperialism: The Biological Expansion of Europe,* pp. 900–1900 (New York: Cambridge University Press, 1990); Donald Worster, *Rivers of Empire: Water, Aridity, and the Growth of the American West* (New York: Oxford University Press, 1985), pp. 3–125.

[19]Lucie Laumonier, A Tale of Plagues, *History Today,* January 1, 2020.

[20]Centers for Disease Control and Prevention, History of the 1918 Flu Pandemic, 2018.

Business as Usual and Plagues

Peter Daszak, a British zoologist with extensive experience in China, is the director of the New York-based non-profit Eco Health Alliance. He has studied the origins of viruses for several years. He was active in the pandemic of 2020–2022. I will say more about that below. In a December 9, 2019 interview,[21] he praised his Chinese colleagues for their outstanding work. Starting at minute 28 of his interview, he said some of the lab-manipulated viruses "can cause SARS disease in humanized mice models [that] … you can't vaccinate against… So, these are a clear and present danger." Listening to this outrageous claim, the interviewer said to Daszak: "You say these are diverse coronaviruses… you can't vaccinate against… and [that there are] no anti-viral [medications effective against them available] – so what do we do?" Daszak's response was unclear. But there's no doubt that he had hinted that coronavirus lab experiments were happening in Wuhan, China, and that they had the potential to lead to a pandemic. Shortly after this interview, the pandemic broke out in Wuhan, China. Daszak probably freaked out. However, he denied that a laboratory accident might have brought on the pandemic.

Daszak turned to nature. He linked development projects and rural people to disease footprints in the natural world. In other words, he blamed loggers and soybean farmers being in places where they should not be; habitats reserved for wildlife, which, when disturbed, facilitate the movement of viruses – like the coronavirus – into human society, in a big way. In an April 16, 2020 interview, he highlighted the anthropogenic origins of the coronavirus pandemic. He ridiculed the idea that the coronavirus was a product of biological warfare. "The idea that this virus escaped from a [Chinese] lab," he said, "is just pure baloney. It's simply not true. I've been working with that lab [in Wuhan, China] for 15 years. And the samples [of bat viruses] collected were collected by me and others in collaboration with our Chinese colleagues. They're some of the best scientists in the world."[22]

---

[21] Peter Daszak interview: *"This Week in Virology," Microbe TV*, December 9, 2019. Also quoted by Nicholas Wade, The origin of COVID: Did people or nature open Pandora's box at Wuhan? *Bulletin of the Atomic Scientists*, May 5, 2021. Wade's lengthy article is full of valuable insights. Another lengthy but less substantive report was published by *Nature* on June 8, 2021: "The COVID lab-leak hypothesis: what scientists do and don't know," by Amy Maxmen and Smriti Mallapaty.

[22] "Pure Baloney": Zoologist Debunks Trump's COVID-19 Origin Theory, Explains Animal-Human Transmission, *Democracy Now*, April 16, 2020.

That may well be true, though the State Department raised serious doubts about the integrity of the Wuhan laboratory: they were found to have failed to have enough skilled technicians to safely carry out the dangerous work it did on bat coronaviruses.[23] Second, Daszak funded and probably drafted a statement by 27 scientists, which denounced anyone suggesting a different origin of the plague, save that of it being a bat virus that originated in the natural world.[24] The British science journal, *The Lancet*, of March 7, 2020, published that statement in which these scientists defended the natural origin of the pandemic and praised the work and dedication of their Chinese colleagues, saying: "We are public health scientists who have closely followed the emergence of [the] 2019 novel coronavirus disease (COVID-19) and are deeply concerned about its impact on global health and wellbeing. We have watched as the scientists, public health professionals, and medical professionals of China, in particular, worked diligently and effectively to rapidly identify the pathogen behind this outbreak, put in place significant measures to reduce its impact, and share their results transparently with the global health community… We sign this statement in solidarity with all scientists and health professionals in China who continue to save lives and protect global health during the challenge of the COVID-19 outbreak. We are all in this together, with our Chinese counterparts in the forefront, against this new viral threat."

Yes, we are in this together, Americans and Chinese. But Daszak and these scientists did not talk about science in their rush to achieve their seemingly political objective, which was to establish that the pandemic had nothing to do with laboratory manipulations of bat viruses or lab accidents. The sole origin of the pandemic was nature.

In an April 16, 2020 discussion Daszak had with Amy Goodman of *Democracy Now*, he prided himself in studying the origins of emerging diseases, saying that about 75% of every new emerging disease originate in wildlife. "Every species of wildlife," he said, "carries viruses that are a natural part of its biology, a bit like we have the common cold and herpes, cold sores." These wildlife viruses are for the most part harmless, though some of them can be lethal to people. There may be as many as 1.7 million unknown viruses in wildlife. Viruses in people are closely related to those

---

[23] Josh Rogin, State Department cables warned of safety issues at Wuhan lab studying bat coronaviruses, *Washington Post*, April 14, 2020.

[24] Sainath Suryanarayanan, EcoHealth Alliance Orchestrated Key Scientists' Statement on "Natural Origin" of SARS-COV-2, *Independent Science News*, November 19, 2020.

from bats. This is not unusual. Bats are full of a variety of viruses...
SARS coronavirus, the original virus, emerged from bats. [The] Ebola
virus is a bat-origin virus. [As are] rabies and many others." Daszak
explained that the huge diversity of bats in Southeast Asia has been a
bridge for the movement of viruses to humans. "[The rural populace] who
live close to bat caves," he said, "cannot escape the bats and their viruses.
Bats fly over them every night. Second, rural people go into the caves of
bats seeking their guano for fertilizer. Most of those people are subsistence farmers who supplement their diet with wildlife, including bats.
They may get exposed to the viruses carried by the bats. In addition, there
is a market for selling both live and dead bats. The Wuhan market in
China has been a great market bringing together lots of people and
wildlife – including bats. That and other markets are really good places
for a virus to spread."[25]

Markets, however, are also very bad places for animals. "The markets
of the Far East," says the famous British naturalist, David Attenborough,[26]
"are notorious. Everybody concerned with animal welfare knows that
these are the hellholes of the natural world... I remember seeing pangolins
in the wet market in Indonesia in 1956. Whether there was a pandemic or
not, there are parts of the natural world where animals are regarded as
objects and treated as though they had no feeling, without any sympathy
of any kind. And it's prevalent all over the world. It's a horrible thing
to see."

Daszak would probably agree with Attenborough. He speaks[27] about
people encroaching into wildlife habitats, opening new roads into forests
for mining and logging and subsistence agriculture. That, he says, represents "a global trend that will drive the rise of future pandemics." He
recommends we rethink our relationship with the natural world, and at
minimum "reduce our ecological footprint." He even turns to "folks on
the right," asking them, "what about your own health? You know, we are
making ourselves sick by making the planet sick." Daszak says he and his
Chinese colleagues found a huge diversity of "bat-origin coronaviruses."

---

[25]"Pure Baloney". Zoologist [Peter Daszak] Debunks Trump's COVID-19 origin theory,
explains animal-human transmission, *Democracy Now*, April 16, 2020.

[26]Etan Smallman, David Attenborough still has hope for our future, *New York Times*,
December 25, 2020.

[27]Unless otherwise indicated, quotes from Daszak come from his interview by Amy
Goodman, *Democracy Now*, April 16, 2020.

The bats keep "spilling [plague viruses] over into people." As a result, about 3% of rural people in Southwest China have antibodies to wildlife viruses. Every year about 1 to 7 million people across Southeast Asia are infected by a zoonotic virus originating from a bat. "So," Daszak says, "it's not just an expectation that we'll have more events. It's a certainty."

Daszak is also concerned about many other viruses he came across in his research in China. He laments we know practically nothing about those viruses. His hope is that we should study them before they make us sick. He insists we should help poor tropical countries deal with the viruses because that way we protect ourselves as well. "It's a right-wing agenda and a left-wing agenda," he says. His most important lesson is that we are to blame for the coronavirus: "It's our everyday way of going about business on the planet that seems to be driving this," he says. Business as usual brings plagues.[28]

The Nature of Pandemics

Peter Daszak represents one point of view in a very controversial political and ecological reality affecting the future of humanity. I agree with him: we should stop invading the natural world for mining, petroleum, logging, and subsistence and commercial cash crop production. He is definitely right in saying we (corporations and governments) are responsible for ecocide and the disease consequences of a wrecked natural world. It is probably true that bats spill viruses over people who intrude too deeply into the remaining fragments of the environment that belong to wildlife. Despite the eminent philosophical, moral, and ecological reasons for the protection of nature from additional human abuse, there are voices connecting the 2020 plague to biological warfare and/or biosafety labs defended by Daszak. There were alleged accusations by certain organizations that Daszak said what he said because of a new grant that his company, Eco Health Alliance, had received in 2020 from the National Institutes of Health (NIH).[29] From 2013 to 2020, the US Agency for International Development (State Department) and the US Department of

---

[28]Mary Harris, A warning from a scientist who saw the Coronavirus coming, *Slate*, March 5, 2020.

[29]Maria Godoy, Group whose NIH grant for virus research was revoked just got a new grant, *National Public Radio*, August 29, 2020.

Defense granted Eco Health Alliance US$39 million in funding. In addition, David Franz, former commander of Fort Detrick, America's biological warfare center, is advising Daszak and Eco Health Alliance.[30]

This military connection is enveloping Daszak with a dark shadow. In addition, he made a religious analogy that is troubling. He imagined there was a connection between the mythical fallen angels and Lucifer of the Book of Revelations to the fallen humans of twenty-first century America punished by plagues. He reached this imaginary conclusion in reference to the 1562 painting, *The Fall of the Rebel Angels, by* Dutch sixteenth-century artist Pieter Bruegel. He described the fall of angels to pathogens striking humans "through an evolutionary (not spiritual) pathway that takes them to a netherworld where they can feed only on our genes, our cells, our flesh. Will we succumb to the multitudinous horde? Are we to be cast downward into chthonic chaos represented here by the heaped up gibbering phantasmagory against which we rail and struggle?"[31] This dark vision overwhelmed Daszak's ethical concerns for the protection of nature and his conviction that our culture is causing pandemics. He rushed to the fashionable but false portrayal of pandemics to terrorism. "Pandemics," he said in early 2020, "are like terrorist attacks: We know roughly where they originate and what's responsible for them, but we don't know exactly when the next one will happen. They need to be handled the same way – by identifying all possible sources and dismantling those before the next pandemic strikes."[32]

Connecting pandemics and terrorism is problematic. Pandemics are not like terrorist attacks. In my opinion, pandemics are anthropogenic ecumenical diseases for the most part, much larger than any terrorist attack. I believe that they are the outcome of bad human actions in the natural world, society and international relations. This toxic legacy came to the surface, once again, after September 11, 2001, when Saudi criminals crashed hijacked planes into the World Trade Centre in New York and the Pentagon in Washington, DC. And now in the third decade of the twenty-first century, the 2020–2022 plague possibly merged bad actions

---

[30] Sam Husseini, Peter Daszak's eco health alliance has hidden almost $ 40 million in pentagon funding and militarized pandemic science, *Independent Science News*, December 16, 2020.

[31] Peter Daszak, A Fall from Grace to… Virulence, *Ecohealth*, March 5, 2008.

[32] Peter Daszak, We knew disease X was coming. It's here now, *New York Times*, February 27, 2020.

in the natural world and bad treatment of domesticated animals into a virulent disease. Certainly, this theory finds support from the destruction of nature, irresponsible biological warfare experiments, and a sick agriculture relying on gigantic quantities of very toxic and disease-causing pesticides and the joint suffering and death of billions of animals.[33] This abuse of nature and agriculture harms people and all animals, including bats.

We are paying the price after centuries of unfortunate European misuse of the natural world in the tropics. The Europeans brought diseases to their colonies. It happened in the United States. The Kiowa of the southern Great Plains of the United States have a myth that explains their dessimation from several epidemics of smallpox brought to their land by Europeans. The legendary hero of the tribe, Saynday, comes across smallpox dressed like a tall white man. They talk to each other and the tall man tells Saynday he is smallpox. "I come from far away, across the Eastern Ocean. I am one with the white men – they are my people as the Kiowas are yours... I bring death. My breath causes children to wither like young plants in the spring snow. I bring destruction. No matter how beautiful a woman is, once she has looked at me she becomes as ugly as death. And to men I bring not death alone but the destruction of their children and the blighting of their wives."[34]

Part of this legacy of exporting pandemics and creating conditions at home for pandemics takes shape in animal farms. I will be supporting this theory below. But before exploring animal farms, let us turn to China.

Wuhan, China

In two reports, one dated April 2020, and the other, August 2021, the US Intelligence Community says the COVID-19 virus was not a biological weapon, that it was not genetically engineered, and that Chinese officials had no knowledge of the virus before the outbreak of the pandemic. The reports proposed that plausible origins for the virus included the transmission of the virus from animals to humans and a possible lab accident.

---

[33] Fowler and Mooney, *Shattering*; Ketcham, *This Land*; USDA, *A Time to Choose*; Krebs, *The Corporate Reapers*; Vallianatos with Jenkins, *Poison Spring*.
[34] Crosby, *Ecological Imperialism*, pp. 207–208.

However, they also emphasized that no strong evidence exists for either of these two possibilities of the origins of the pandemic.[35]

Another narrative suggested by Peter Daszak, cited above, argues that the virus that caused the pandemic possibly jumped from bats to humans in the Huanan Seafood Wholesale Market, a wild animals bazaar selling dead and live bats in Wuhan, China. The followers of this speculation believe that infected Chinese then traveled to other cities in China, the United States, and all over the world, spreading the coronavirus and making it a pandemic. Wuhan is the capital of Hubei province. It is also a gigantic city hosting the Wuhan Institute of Virology and the Wuhan Center for Disease Control and Prevention. Both of these scientific centers study coronaviruses. The Wuhan Institute of Virology sends teams of scientists in search of bats in the forests near Wuhan. These experts enter bat caves and catch bats for the study of their viruses. This is a hazardous task as some of the viruses are very dangerous. Lab scientists extract two or more viruses from captured bats. Out of these viruses, they engineer a hybrid species that they call the Chimera virus. This Chimera virus could be pathogenic enough to cause a pandemic.[36] The name "Chimera" comes from Greek mythology about a monster made up of three animals: a lion with the neck and head of a goat sprouting from its back and a snake head at the end of the lion's tail. Are biological engineers fiddling with viruses in laboratories all over the world on their way to engineering another Chimera monster? Or, as Daszak said, are they studying the viruses to avoid their diseases? Virologists probably are aware that no one can control a lab-engineered Chimera monster. So, why are they pursuing such experiments? In theory, they may say that they are trying to develop vaccines against the coming plagues from biological warfare research lab accidents and continuing ecocide.

We know that some of the research at the Wuhan Institute of Virology were funded by the governments of China and America. Peter Daszak was part of that US funding and influence peddling in Chinese virology labs. Richard H. Ebright, professor of chemistry and chemical biology from Waksman Institute of Microbiology, Rutgers University, reported that the Chinese and American scientists working together at the Wuhan Institute

---

[35] Office of the Director of National Intelligence, Intelligence Community Statement on Origins of COVID-19, Press Release, April 30, 2020; Covid-19 Origins, August 2021.

[36] Filippa Lentzos, Natural spillover or research lab leak? *Bulletin of the Atomic Scientists,* May 1, 2020.

of Virology had been engineering Chimeras in order to study the degree to which bat viruses could infect humans. Their genetic engineering methods left no fingerprints, however. He explained:

"[Chinese and American scientists at the Wuhan Institute of Virology] constructed a series of novel chimeric viruses encoding different receptor binding domains – with different receptor binding affinities – in an otherwise constant genomic context. And did so using 'seamless ligation' procedures that leave no signatures of human manipulation."[37] In short, Professor Ebright was saying that American and Chinese scientists at Wuhan are so good in the genetic engineering of bat viruses that they constructed Chimeras with pandemic disease potential so flawlessly that no one could backtrace their handiwork.

It is not clear if there is a connection between the outbreak of the virus pandemic in Wuhan and the Wuhan biolab. But what is clear is that on the eve of the lockdown of Wuhan, on January 23, 2020, five million Chinese from Wuhan left the city. They visited relatives in thousands of Chinese cities and traveled abroad.[38]

Biological Warfare Labs

The other more problematic theory is that biological warfare labs engineered the coronavirus into a deadly species. This engineered Chimera virus then either escaped from the lab by accident or was released intentionally as a weapon of war. Accidental lab releases of pathogens are not that unusual.[39] Federal authorities recorded more than 1100 lab accidents between 2008 and 2013. These so-called "mishaps" had to do with releases or spills of viruses, bacteria and poisons of significant bioterror risks to public health and agriculture.[40] However, neither China nor America would have willfully started a biological war for a simple reason: such a war, like nuclear conflicts, would have no winners. Thus, if this

---

[37] Tweet, May 16, 2020.

[38] Wang Xiuying, The Word from Wuhan, *London Review of Books*, March 5, 2020. See also: Alice Su, You don't know that feeling of terror: Wuhan's survivors find no closure from the coronavirus, *Los Angeles Times*, May 24, 2020.

[39] Sam Husseini, The long history of accidental laboratory releases of potential pandemic pathogens, *Independent Science News*, May 5, 2020.

[40] Alison Young, Hundreds of bioterror lab mishaps cloaked in secrecy, *USA Today*, August 17, 2014.

coronavirus was an engineered Chimera pathogen, it might have escaped a lab by mistake.

Biological Warfare

The second and equally plausible avenue is biological warfare. The question is, why would highly educated humans be risking civilization? They know or should know the history of the inhumane and holocaust-like experience of the chemical warfare employed during WWI as well as the mayhem and consequences of the first use of atomic bombs over Japan towards the end of World War II. Are they indulging in games of biological warfare? Is humanity their target? These biological experts should have known that genetic engineering, the mother of biological warfare, could potentially become the "nuclear bomb" of poor nations.

It is not that these biological warfare enthusiasts are living in a vacuum. Some scientists are saying certain things should never be done, or that fellow researchers should, at least, rethink the process and stand on the side of public health. In 2014, two such scientists, Marc Lipsitch and Thomas Inglesby, recommended a moratorium for the aggressive biology of the weaponization of microbes.[41] They said that research designed "to create new potential pandemic pathogens – novel microbes that combine likely human virulence with likely efficient transmission in humans… poses extraordinary potential risks to the public." In 2023, two other scientists, Marc Siegel and Robert Redfield, urged a pause on manipulating viruses.[42] They accused NIH of playing games with allowing and not allowing gain of function (weaponization) of viruses. "Science," they said, "should always be guided and restricted by strict moral and ethical principles. We are calling for a moratorium on all gain of function research, where it is forbidden to deliberately alter a pathogen to provoke or assess its ability to spread among or sicken humans. This applies to research going on in pharmaceutical companies, in universities and anywhere else. If gain of function did not cause this ferocious pandemic, it most certainly can cause the next one."

---

[41] Marc Lipsitch and Thomas Inglesby, Moratorium on research intended to create novel potential pandemic pathogens, *MBio*, December 12, 2014.

[42] Marc Siegel and Robert Redfield, To prevent a deadlier pandemic, pause gain of function research, *The Hill*, February 2, 2023.

The novelist Nicholson Baker agreed with such a moratorium. "We need to hear from the people who, for years, have contended that certain types of virus experimentation might lead to a disastrous pandemic like this one [the COVID-19 pandemic]. And we need to stop hunting for new exotic diseases in the wild, shipping them back to laboratories, and hot-wiring their genomes to prove how dangerous to human life they might become."[43]

Richard Ebright has had similar anxieties and concerns. He exposed the hypocrisy of the American funders of the Chinese virologists who practiced the genetic engineering of viruses that the Americans taught them. In a March 24, 2021 interview,[44] he shook his finger at the guilty parties, saying: "The Director of the National Institute of Allergy and Infectious Diseases [(NIAID), Anthony Fauci, retired] and the Director of the National Institutes of Health [(NIH), Francis Collins, retired in 2021] have systematically thwarted efforts by the White House, the Congress, scientists, and science policy specialists to regulate GoF [growth of function] research of concern and even to require risk-benefit review[s] for projects involving GoF research of concern... In 2014, the Obama White House implemented a 'Pause' in federal funding for GoF research of concern. However, the document announcing the Pause stated in a footnote that: 'An exception from pause may be obtained if head of funding agency determines research is urgently necessary to protect public health or national security.' Unfortunately, the NIAID Director and the NIH Director exploited this loophole to issue exemptions to projects subject to the Pause – preposterously asserting the exempted research was 'urgently necessary to protect public health or national security' – thereby nullifying the Pause."

In the same spirit, but more diplomatically than Ebright, the science writer Nicholas Wade blames virologists for converting science into biological weapons while maintaining spy-like devotion to silence and funders. He did this in a fascinating and extremely long article published in the *Bulletin of the Atomic Scientists* on May 5, 2021. He chastised the US government for funding the Wuhan Institute of Virology to create Chimeric coronavirus monsters. He was equally angry at the Chinese

---

[43] Nicholson Baker, The Lab-Leak Hypothesis, *New York Magazine*, January 4, 2021.
[44] Jorge Casesmeiro Roger, An Interview with Richard Ebright: The WHO [World Health Organization] Investigation [of the coronavirus pandemic] Members were "participants in disinformation," *Independent Science News*, March 24, 2021.

scientists for doing biological warfare experiments. And he did not like the lockdown of the laboratory by the Chinese government. Wade seems to be blaming the Chinese for embarking on the dangerous experiment of adding pandemic disease power and contagion to the coronaviruses they extracted from bats. Wade is one of the few reporters who, at least, shares the blame for the weaponization of the coronavirus between America and China; Yes, the Chinese Wuhan Institute of Virology used genetic engineering to turn bat viruses into potential biological weapons, but Americans are equally to blame for training and funding Chinese scientists to do exactly what they did. Nevertheless, Wade argued that no real evidence existed in 2021 that the coronavirus escaped from the laboratory.

In a September 23, 2021 article,[45] Wade repeated his earlier argument about the American–Chinese collaboration at the Wuhan Institute of Virology in engineering viruses of pandemic virulence. He was bewildered with the silence and inactivity of America's eighteen intelligence agencies. Where were those thousands of spies while Chinese and American virologists were engineering the SARS-CoV-2 virus? He criticized the Trump administration's Secretary of Health and Human Services, Alex Azar, and the director of the NIAID, Anthony Fauci, for failing to supervise Daszak. This "eco-health" entrepreneur paid himself a salary of about US\$411,000 a year from his Pentagon contracts, aiming to "defuse the potential for spillover of novel bat-origin high-zoonotic risk SARS-related coronavirus in Asia."

The US government spent about US\$125 million studying bat viruses in Asia, Africa, and Latin America. In other words, there was lots of money for diverting attention from the hazardous lab experiments with bat viruses. Wade was angry with Daszak because of his "cowboy schemes to have his Chinese subcontractors generate ever more dangerous viruses in minimally adequate safety conditions."[46] A Subcommittee of the US House of Representatives accused Daszak and his Eco Health Alliance for acting with "contempt for the American people."[47] By 2024, Wade also changed his mind and accused the Chinese of letting the engineered virus

---

[45] Nicholas Wade, New Routes to Making Covid-19 in the Lab, Nicholaswade.medium.com, September 23, 2021.

[46] *Ibid.*

[47] Select Subcommittee on the Coronavirus Pandemic, An evaluation of the evidence surrounding eco health alliance research activities, Interim Staff Report, May 1, 2024.

escape from the Wuhan Institute of Virology.[48] Another scientist, Alina Chan, blasted the institute in China for sparking the pandemic. "A laboratory accident," she said, "is the most parsimonious explanation of how the pandemic began."[49]

Not all American scientists who worked with their Chinese colleagues accused them of causing the pandemic. One of those American scientist was diplomatic. He concerned himself about the integrity of the Wuhan virus experiments. This was James Le Duc, professor of microbiology and former director of Galveston National Laboratory at the University of Texas Medical Branch. Le Duc was one of the American scientists who trained Chinese microbiologists at the Wuhan Institute of Virology. He wanted his Chinese colleagues to be proactive and investigate their own work so as to discover and, promptly correct, any mishaps which could have compromised the safety of their laboratory or led to the accidental rumored leak of the coronavirus. On February 9, 2020, he sent an email to his Chinese professor friend, Yuan Zhiming, at the Wuhan Institute of Virology. He urged him to "aggressively address... rumors [of a lab leak] and presumably false accusations quickly and provide definitive, honest information to counter misinformation. If there are weaknesses in your program, now is the time to admit them and get them corrected. I trust that you will take my suggestions in the spirit of one friend trying to help another during a very difficult time." Professor Zhiming did not bother to respond to his message. This probably infuriated Le Duc enough that, two months later, in April 2020, he changed his opinion about Wuhan. He wrote to Philip Russell, former president of the American Society of Tropical Medicine and Hygiene, that it was "certainly possible a lab accident was the source of the epidemic." He added, "we can't trust the Chinese government."[50]

In contrast to Le Duc and other observers, Daszak persisted in his defense of the scientific integrity of the Wuhan Institute virologists.[51] But Daszak is not alone in wishing for people to believe that the spillover of

---

[48] Nicholas Wade, Story of the decade, *City Journal*, January 25, 2024.
[49] Alina Chan, Why the pandemic probably started in a lab, *New York Times*, June 3, 2024.
[50] Shannon Murray, Biosafety expert close to Wuhan Institute of Virology urged associates there to address his tough questions about lab origins of SARS-CoV-2, *US Right to Know*, December 2, 2021.
[51] Jimmy Tobias, During a heated covid origins hearing, a scientist comes in for questioning, *The Nation*, May 3, 2024.

wild bat viruses brought about the pandemic. Researchers studying bat viruses from Laos, Cambodia, and Central China concluded that those viruses have a molecular hook similar to the hook of the wild viruses that "caused" the 2020 pandemic. In either case, the virus hook facilitates the virus' jump onto human cells.[52]

Nature's Revenge

Bats have become a potential weapon in the foolish games of genetic engineering, biological warfare and biodefense. Bats have been hunted, killed and eaten by humans for centuries. Humans harming and killing bats ignore the fact that the bats are extremely beneficial animals. Apart from only flying at night to look for food, some species of bats propagate the seeds of valuable plants and trees. Others pollinate plants and crops. But all bats protect the natural world and us with their voracious eating of insects – disease-spreading mosquitoes in particular. Yet in the citation below, David Quammen argued that the destruction of forests have caused about 200 species of bats to lose their natural habitats, forcing them into closer proximity with humans. Bats may be "spilling" the viruses to humans, but not because they don't like us; it is because humans are getting too close to their caves. And in the last fourteen years in North America, humans have become a plague for these "majestic creatures," which are dying at a "cataclysmic rate" from a contagious disease (white-nose syndrome) we are spreading among them.[53]

The plight of bats is a mirror of wildlife pushed to potential decimation and eventual extinction by humans in a frenzy. We are acting like we came from outer space, invading and conquering this beautiful planet; home to people, science, civilization, and myriad forms of life for millennia. My hunch is that the virus plague of 2020–2022 was a culmination of viruses resulting from the devastation of nature, biological warfare lab accidents, and horrific animal farms. The key ingredients in this development include mining, fossil fuel extraction, burning of gigantic forests like those of Africa, the Amazon, Australia, Canada and California, logging, industrialized agriculture, countless millions of cattle grazing grasslands,

---

[52] Carl Zimmer, New discovered bat viruses give hints to Covid's origins, *New York Times*, October 14, 2021.

[53] David Quammen, A few words on behalf of bats, *New York Times*, December 13, 2020.

billions of petroleum-burning cars, trucks and leaf blowers, billions of gas-burning homes, and the wanton destruction and pollution of the natural world on land and in the seas. Slowing down or reversing this large-scale pollution would be a fantastic accomplishment. It will be difficult, challenging our political institutions to the outmost. Yet science demands radical changes and the rethinking of our doings, especially those activities that come under the umbrella of economy and business, if only to stop widespread destruction of nature and, in good time, our potential death and possible extinction.

In 1984, Hugh Iltis, the fearless professor of botany at the University of Wisconsin, equated "economic development" to cancer feeding on "biotic destruction."[54] A few years later, in the early 1990s, Robert Ayres and Udo Simonis, scientists of the United Nations University, described the global spread of industrialization during the last two centuries as a cancer. "[I]ndustrialization, in its present form," they said, "is a process of uncontrolled, unsustainable 'growth' that eventually destroys its host – the biosphere."[55] In the late 1990s, British biologists warned that, since the 1960s, the industrialization of agriculture in Western Europe had devastating effects on nature. Research evidence cited below, says that one-fifth of Europe's birds are endangered and threatened. The overall decline of birds, insects, and plants in northern Europe was quite dramatic in the last thirty years of the twentieth century. This was a time when the homogenization of Europe's countryside was pursued with rigor and "science." European farmers, earning huge subsidies, abandoned their centuries-old agrarian traditions and demolished their beautiful small-farm landscape for the icon of the plantation, growing rarely more than one crop. In England, between the years 1968 to 1995, farm birds like the skylark and corn bunting declined by 30%.[56]

---

[54] Hugh H. Iltis, The Extinction of Life on Earth: Asking the Proper Questions in *Threats to the Tropical Biota* (Symposium on the Biogeography of Mesoamerica, Merida, Yucatan, Mexico, 26–30 October 1984).

[55] Robert U. Ayres and Udo E. Simonis, eds., *Industrial Metabolism: Restructuring for Sustainable Development* (Tokyo: United Nations University Press, 1994), p. xii. See also Robert N. Proctor, *Cancer Wars: How Politics Shapes What We Know and Don't Know About Cancer* (New York: Basic Books, 1995; Samuel Epstein, Winning the war against cancer?... Are they even fighting it? *The Ecologist*, March/April 1998, pp. 69–80.

[56] John R. Krebs *et al.,* The second silent spring? *Nature*, August 12, 1999, pp. 611–612.

The United Nations Environment Program issued a similar dire alarm in 2001[57]: "Human activities are destroying [the] Earth's biological wealth at an unprecedented rate. There is a strong consensus that the extinction spasm now being caused by human activities is greater than any [extinction] since the dinosaurs died out 65 million years ago. This damage is irreversible and – many believe – unethical. What's more, given humanity's dependence on food crops and other biological resources, it is also dangerous to our species." The World Wildlife Fund, a global civil society organization, reached similar conclusions. In its 2002 Living Planet report, WWF confirmed that humanity's onslaught against nature was taking a terrible toll. In the relatively short span of time between 1970 and 2000, the world's natural ecosystems declined by about 37%. This translates into a loss of 15% of the terrestrial species, the destruction of some 35% of all animals and plants living in seas and oceans, and the disappearance of about 54% of all species of animals and plants living in freshwater.

Gerardo Ceballos (National Autonomous University of Mexico), Paul R. Ehrlich and Rodolfo Dirzo (Stanford University), distinguished biologists, spoke of the severity of the ongoing sixth mass extinction that is wiping out countless vertebrate animals (mammals, birds, reptiles, and amphibians) and plants. Indeed, they emphasized the seriousness of the destruction of animals going on all over the world by speaking about the population extinction pulse that is the biological annihilation of vertebrate animals caused by human activities. These economic activities include taking forests and land away from animals for plantations and fracking the land for petroleum and natural gas as well as for golf courses and mega cities; the overexploitation of land, forests, mountains, deserts; and the ceaseless poisoning of the environment through various human activities. Diseases, human overpopulation, continuing population growth, overconsumption, particularly by the rich, and a probable "large-scale nuclear war" also threaten the natural world.[58]

---

[57] UN Environment Programme, Sustainable Agri-Food Production and Consumption Forum: Key Issues and Information Sources – Agri-Food Production and Biodiversity, (2001). http://www.agrifood-forum.net/issues/production.htm.

[58] Gerardo Caballos, Paul R. Ehrlich and Rodolfo Dirzo, Biological annihilation via the ongoing sixth mass extinction signaled by vertebrate population losses and decline, *Proceedings of the National Academy of Sciences*, July 25, 2017.

In my opinion, overfishing is yet another blow against a healthy planet. It impoverishes the seas and oceans, threatening marine ecosystems and the food security of dozens of countries. For example, in the last fifty years, commercial fishermen have caused a worldwide staggering loss of sharks and rays. These valuable and iconic species declined by 71%.[59]

Paul Ehrlich and other researchers warn that the planetary destruction of vertebrate animals "will itself promote cascading catastrophic effects on ecosystems, worsening the annihilation of nature... while the biosphere is undergoing mass species extinction, it is also being ravaged by a much more serious and rapid wave of population declines and extinctions... Humanity will eventually pay a very high price for the decimation of the only assemblage of life that we know of in the universe."[60]

These reasons explain the idiocy of the political controversies in so many countries, which fail to give cover to the pollution of the world, indeed, the undermining of the natural world and extinction of so many species on this beautiful Earth. Pesticides and one-crop farming have been decimating healthy, threatened, and endangered species.[61] Industrialized agriculture is probably the largest trigger for the sixth mass extinction.[62] The irreversible effects of the sixth mass extinction is equivalent to dramatic and tragic "population declines and extirpations, which will have negative cascading consequences on ecosystem functioning and services vital to sustaining civilization."[63]

In 2020, the World Wildlife Fund International issued another dire warning.[64] According to them, "[N]ature is unravelling and... our planet

---

[59] Catrin Einhorn, Shark populations are crashing, with a very small window to avert disaster, *New York Times*, January 27, 2021.

[60] Gerardo Caballos, Paul R. Ehrlich and Rodolfo Dirzo, Biological annihilation via the ongoing sixth mass extinction signaled by vertebrate population losses and decline, *Proceedings of the National Academy of Sciences*, July 25, 2017.

[61] Center for Biological Diversity, Landmark lawsuit filled to protect hundreds of rare species from pesticides, January 20, 2011.

[62] Rosemary Mason, The sixth mass extinction and chemicals in the environment: Our environmental deficit is beyond nature's ability to regenerate, *Journal of Biological Physics and Chemistry*, 15 (2015), pp. 160–176.

[63] Gerardo Ceballos *et al.*, Biological annihilation via the ongoing sixth mass extinction signaled by vertebrate population losses and declines, *Proceedings of the National Academy of Science of the United States*, July 25, 2017.

[64] WWF International, *Living Planet Report, 2020*.

is flashing red warning signs. Humanity's destruction of nature is having catastrophic impacts not only on wildlife populations but also on human health and all aspects of our lives.... Nature is declining globally at rates unprecedented in millions of years. The way we produce and consume food and energy, and the blatant disregard for the environment entrenched in our current economic model, has pushed the natural world to its limits.... It's time for the world to agree [to] a New Deal for Nature and People, committing to stop and reverse the loss of nature by 2030 and build a carbon-neutral and nature-positive society. This is our best safeguard for human health and livelihoods in the long term... to ensure a safe future for our children."

Yes, a New Deal for Nature and People would be the right thing to do. But petroleum complicates everything. Billionaires see trillions of dollars locked in the buried oil, climate change or not. The coronavirus did not disrupt their power, though its disruption (economic, social and political) was large and global. In the words of journalist Bill McKibben, the destruction of the Earth is "hardwired" into those who run the world – oil billionaires in particular. These billionaires are part of the "systems" McKibben refers to when he says, "only by attacking those systems... ripping out the fossil-fueled guts and replac[ing] them with renewable energy, even as we make them more efficient, can we push emissions down to where we stand a chance."[65]

The next chapter examines further the origins of the 2020–2022 Covid pandemic. Plagues have been accompanying humans for millennia. In each case, humans did spark the plague. The same thing happened in 2020, though the destruction of nature has been unprecedented in the last few decades.

---

[65] Bill McKibben, 130 Degrees, *The New York Review*, August 20, 2020.

5 Plague Mills

An animal farm in Central Valley, California.
Photo by Evaggelos Vallianatos.

Prologue

Decades-long development in and near tropical forests has brought humans closer to bat habitats, often caves. Caves for bats and other unsuspecting wildlife sanctuaries are now being visited by trappers, hunters, wildlife merchants and scientists. Bats and humans do not get along. "They [bats] have been calumniated and abused for centuries," says the zoologist David Quammen. "Some people, mainly from the comfort of distance and ignorance, find bats repellent and spooky. Some people fear them, with or without rational grounds. Bats are sometimes slaughtered in large numbers, defenseless at their collective roosts, when people deem them menacing, inconvenient, noxious or desirable as food."[1]

Bats are also loaded with viruses, which are harmless to them. Wang Linfa, Director of the Emerging Infectious Diseases Program at Duke–NUS Medical School and Chairman of the Scientific Advisory Board of the Wuhan Institute of Virology, China, has a tremendous respect for bats. "Bats," he said, "are resilient to viruses that can kill humans. If we can learn from bats to do what they do, then we will be very fortunate."[2] Those viruses have probably coevolved and coexisted with bats and other wildlife, even though no scientific studies are available. Therefore, human intrusion into the relationship between bats and their viruses comes at heavy costs.

Some blame Chinese traditional medicine for encouraging the eating of rare, wild animals for wishful therapeutic and "miraculous" effects.[3] While these beliefs shed some light on Wuhan and its wildlife market, it does not explain why the virus explosion took place in 2020 and not before. There were many arguments on whether the COVID-19 pandemic may or may not have been a lab accident. I theorize that like past global epidemics, it served as a warning that humans were harming the natural world. Outrageous human development activities change the climate, warming and threatening planet Earth. And nature (the Earth) is probably fighting back. In my opinion, climate change is causing chaos with floods, fires, storms, hurricanes, droughts, and heat waves. These catastrophic effects are seemingly weakening human and environmental health, which

---

[1] David Quammen, The virus, the bats and us, *New York Times*, December 11, 2020.
[2] Hannah Beech, Amid a pandemic, "Batman" matters more than ever, *New York Times*, June 12, 2020.
[3] Wang Xiuying, The Word from Wuhan, *London Review of Books*, March 5, 2020.

may cause accelerating pandemic diseases, and, in theory, may exacerbate the potential dire consequences of the climate chaos enveloping the planet and humanity.

I have repeatedly argued that these joined crises, climate chaos and the pandemic, are the consequences of human greed. In this chapter, I would add the argument that large farmers are equally responsible for boosting climate chaos and potentially introducing diseases that have the capability to become pandemics. They have built such gigantic enclosures for their countless millions of food animals that those facilities can possibly become plague factories if a disease breaks out. Metaphorically speaking, we can say that these animal farms have become shops for cattle, chicken, and pigs. Taking the cue from my 25 years of experience in the US EPA, I believe that feeding these animals corn and soybeans mixed with antibiotic drugs is a temporary but determined measure of agribusiness trying to feed the world.

The COVID-19 pandemic could have become much larger and deadlier. What I mean to say is, if we continue business as usual, the next pandemic may be more destructive than the 2020–2022 global disease. Yet we know little about viruses, while we exploit the natural environment around these microscopic organisms that may transform into dangerous diseases. Related to that, we are woefully ignorant and apathetic to the destruction of natural habitats. Billionaires and corporations around the world, with the sanctions of their governments, have been destroying the natural world,[4] our sole source of life. In my opinion, both industrialized and industrializing economies worldwide are contributing to climate change and natural habitat loss, resulting in the higher risk of plagues.

Caging Animals

I have visited animal farms and saw how animals are caged next to each other in small enclosures. Describing the suffering and violence against factory raised and slaughtered animals is not easy. "How we treat farm animals today," says *New York Times* Opinion Columnist Ezra Klein,

---

[4]Alfred W. Crosby, *Ecological Imperialism* (Cambridge: Cambridge University Press, 2004), pp. 195–216.

"will be seen, I believe, as a defining moral failing of our age." He said that the experience is "mind-melting."[5] It is also barbaric.

Since the caged animals can only eat, they become obsessed with food. They also get sick from a variety of diseases; we have seen chicken developing the deadly H5N1 virus (commonly referred to as "Bird Flu") and pigs, the H1N1 virus ("swine flu") – both of which have been known to also infect humans. In countries like England and the United States, agribusiness owners try to control these diseases by feeding drugs to the animals, especially feeding cattle "indiscriminate" amounts of antibiotics,[6] which, in my opinion, may have threatened the efficacy of antibiotics for human beings. The swine flu was first discovered in Mexico. The bird flu appeared in China, Kazakhstan, Mongolia, Cambodia, Indonesia, Turkey, Egypt, Thailand and Vietnam. Since 2003, the bird flu harmed the poultry industry in Asia and Europe. In 2011, the UN FAO reported that Bangladesh, China, Egypt, India, Indonesia and Vietnam were regularly affected by the bird flu.[7,8] The bird flu also reached Europe and America. "Between 16 March and 14 June 2024," reported the European Center for Disease Prevention and Control, "42 highly pathogenic avian influenza (HPAI) A(H5) virus detections were reported in domestic (15) and wild (27) birds across 13 countries in Europe."[9] The bird flu, avian influenza A (H5N1), says Yale Medicine, "has killed millions of wild birds, as well as caused sporadic outbreaks among poultry and an ongoing multistate outbreak among cows in the United States."[10] European authorities suggest that environmental contamination causes bird flu.[11]

---

[5] Ezra Klein, We will look back on this age of cruelty to animals in horror, *New York Times*, December 16, 2021.

[6] Eric Schlosser, *Fast Food Nation: The Dark Side of the All-American Meal* (Boston: Houghton Mifflin, 2001), p. 221.

[7] US Centers for Disease Control and Prevention, 2010–2019 Highlights in the History of Avian Influenza (Bird Flu) Timeline.

[8] L. D. Sims, *et al.*, Origin and evolution of highly pathogenic H5N1 avian influenza in Asia, *Vet Record*, August 6, 2005, **157** (6), pp. 159–164. https://doi.org/10.1136/vr.157.6.159.

[9] European Center for Disease and Prevention and Control, Avian Influenza Overview, March–June 2024.

[10] Yale Medicine, September 10, 2024.

[11] European Center for Disease Prevention and Control, Facts about avian influenza in humans, February 8, 2023.

We have already touched on the issue of mechanical and chemical agribusiness and why its animal factories are bad for animals, humans, and climate change. Animals are also kept in large and small cages, depending on their sizes. These crowded animal farms are often found in the midst of communities experiencing incredible poverty, where the living conditions for the humans are less than desirable. With lower standards of hygiene, the living conditions for both the animals and humans are a probable environment for disease. Under such circumstances, some of these diseases may move to wild, migratory birds, which have the potential of spreading it beyond Asia.

Governments fail to investigate animal farms and industrialized agriculture as sources of disease. Instead, they resolved animal diseases by culling all animals in the infected area and quarantining the rest in regions next to the affected ones. Some of these culling were shocking due to the high numbers of animals killed, for example, the culling of about 17 million minks in Denmark alone in 2020.[12] This killing rationale has cost the lives of hundreds of millions of domesticated animals.

Then in the 1980s and 1990s, Bovine Spongiform Encephalopathy (BSE), commonly known as "mad cow disease," struck the cattle of 24 countries and especially England, starting in 1986. Mad cow disease is a fatal brain disease caused by a prion. Prions are misfolded infectious proteins. Feeding cattle the meat of other cattle tainted by a prion causes BSE. People eating meat infected by BSE risk contracting the fatal brain disease known as variant Creutzfeldt-Jacob disease (vCJD).[13]

With some exceptions, international institutions like the World Health Organization (WHO) and the Food and Agriculture Organization (FAO) are telling farmers all over the world to get big or get out, exactly what the US government has been telling farmers in America for decades. The advocates of this axiom failed to recognize that large animal farms are not the "solution to a problem they helped create."[14] No wonder these UN institutions remain silent about the agribusiness origins of the global pandemic disaster, COVID-19. In addition, I surmise that governments and international agencies use the endless crises of agribusiness, including

---

[12]Adrienne Murray, Denmark shaken by culling of millions of mink, *BBC News*, November 11, 2020.

[13]Classic Creutzfeldt-Jacob Disease, May 13, 2024.

[14]Lucile Leclair, The biosecurity myth that is destroying small farming, *Independent Science News*, November 10, 2020.

the COVID-19 pandemic and biosecurity, to undermine small family farming in the industrialized world and the traditional farming of the people in the tropics.[15]

Contracting Disease

It has been documented in various studies that industrial animal farms – indeed, the entire fabric of agribusiness in my opinion – not just farmers raising chicken and pigs in the countryside, have been a major source of diseases. They rely on pesticides, which are designed to kill outright or give diseases to living organisms and plants, including humans. Industrialized agriculture "is a potential driver of human infectious diseases."[16] Animal farms add to the burden of diseases related to agriculture. Zoonotic or animal diseases increase with deforestation for pasturing animals. Growing food leaves an ecological footprint "before and after the farm gate."[17] In addition, animal farms generate huge amounts of manure that become the breeding ground for viruses with the potential of exploding into pandemics – the main thesis of this book.

Butchering animals from animal farms may contaminate the meat with bacteria from the intestines and feces of the animals. These bacteria may end up consistently contaminating the feed produced in these farms. Agribusiness merchants are selling feed to farmers and animal farms, which may accidentally be contaminated with feces and diseased animal flesh. Moreover, animal feed is sometimes tainted by toxic chemicals like dioxin.[18] In addition, animal factory farms supervise "contract" farmers in the industrialized world and, sporadically, farmers in the tropics raise animals for the market. Agribusinesses provide the chicks and the piglets to their "contract" workers. Those animals may carry latent viruses. I believe that we could potentially reduce the severe impact of the swine flu if we addressed the agribusiness conditions that contribute to it.

---

[15] Vallianatos, *Fear in the Countryside*.

[16] Hiral A. Shah *et al.*, Agricultural land-uses consistently exacerbate infectioud diseases, *Nature*, September 20, 2019.

[17] Matthew N. Hayek, The infectious disease trap of animal agriculture, *Science*, November 2, 2022.

[18] Judy Dempsey, Tainted animal feed found in France and Denmark, *New York Times*, January 11, 2011.

Above all, we must resolve the issue of the crowded, unhygienic, and inhumane environments of the animal factories around the world.

Trenches of War

We have already mentioned the dreadful conditions of World War I, which contributed to the influenza pandemic of 1918. The virus was born and spread in the trenches of the Western front. I believe that a similar pandemic like that is unlikely, unless we continue the globalization of animal farms. I take that intermingling billions of potentially disease-ridden pigs with the Western-inspired technologies of animal factories in Latin America, Asia, Africa, and Eastern Europe could potentially cause another pandemic. It is also possible that the antibiotics of the high-tech hog and chicken factories of North America and Europe will not suffice for long in preventing the onslaught of the swine and bird flu.[19,20] It would appear that evolution is on the side of the submerged virus. Hog and chicken factories may have been manufacturing their own lethal viruses, which could mutate into an influenza epidemic or even a pandemic.

Pandemic Chaos

In 2021, Gary Ruskin of the US non-profit organization "Right to Know" wrote, "a fearful world was struggling to emerge from a paralyzing pandemic, a confusing health care crisis that emerged swiftly to sicken and kill millions. Today, we are still struggling to find our way back from the catastrophic global consequences of the vicious coronavirus. And we are still without answers as to how and why this virus emerged seemingly out of nowhere."[21] Of course, the pandemic must have emerged from somewhere. We just do not have reliable answers yet. As I have suggested in previous chapters, those who might be responsible for the pandemic are protecting themselves. The effects of the COVID-19 pandemic were immense. The Omicron variant out of South Africa swiftly captured

---

[19]Vallianatos, We must reclaim our farmland from the rural oligarchy, *Truthout*, May 20, 2013.

[20]Vallianatos, Brewing pandemic pestilence, *Counterpunch*, September 6, 2022.

[21]Gary Ruskin, Covid-19 information, *US Right to Know*, November 9, 2021.

Europe and America. The US Centers for Disease Control[22] said that during the second and third week of December 2021, new infections from omicron rose from 13 to 73%. Due to its highly contagious nature, the coronavirus pandemic disrupted societies severely. It imperiled human beings all over the planet, killing close to a million Americans and 14.9 millions worldwide.[23] The virus was invisible and could be everywhere. In 2020 and 2021, when you turned on the radio or television, the primary conversation for those two years was about this invisible and incomprehensive enemy. In late December 2021, the country was facing a "formidable winter."[24] In 2020, restaurants closed. Even factories and airports shut down. Schools closed. Libraries closed. The turmoil of the pandemic highlighted the fragility of students. Isolation, anxiety, zoom teaching, depression, and not a few suicides threatened colleges across the country.[25]

In 2020, Adam Westbrook, an Emmy-nominated filmmaker, journalist and video producer, together with Sanya Dosani, a documentary filmmaker and producer, alleged that Trump lied to Americans repeatedly about the pandemic.[26] It was also said that congressmen, senators and Trump were "fighting" the virus with tons of money, something like two trillion dollars[27] as a stimulus to the economy and individual Americans in response to the various difficulties faced by individual Americans and crucial corporations such as airlines, hospitals, hotels, and public service corporations. Retired general Barry McCaffrey said it best. He denounced the "Revolting sycophancy of [vice-president] Pence and others in the [Trump] Administration... There are eerie echoes of 'supreme leader' adulation to all of this. That Trump tolerates or needs this kind of faux devotion is dangerous in a democracy."[28]

---

[22] Matt Field, Do we need a new vaccine for omicron? Fauci says not yet, other experts disagree, *Bulletin of the Atomic Scientists*, December 21, 2021.

[23] United Nations, Department of Economic and Social Affairs, 2022.

[24] Rong-Gong Lin and Emily Alpert, A perfect storm for Delta, Omicron could overwhelm hospitals within weeks, *Los Angeles Times*, December 17, 2021.

[25] Anemona Hartocollis, Another surge in the virus has colleges fearing a mental health crisis, *New York Times*, December 22, 2021.

[26] Adam Westbrook and Sanya Dosani, The president is lying about coronavirus, *New York Times*, March 18, 2020.

[27] Claire Foran, *et al.,* Trump signs historic $ 2 trillion stimulus, *CNN*, March 27, 2020.

[28] X (formerly Twitter), Barry McCaffrey, March 14, 2020.

Ignorance and Cruelty

On March 24, 2020, Jeffrey Sachs,[29] professor of economics at Columbia University in New York, denounced America's "health system" as one that primarily focused on "making money." He bemoaned giving "monopoly power to powerful companies, who then use their unbelievable profits, in part, to buy the Congress." "Corruption," Sachs said, "of our political system has driven so much attention to the wrong things, away from our well-being and now even away from our survival." To this political decline, he added, the mantra of conservatives is to stop spending money for saving lives but, instead, spend money to protect the economy. "This is," he said, "a corruption of the most basic human spirit. It's a kind of sickness that has infiltrated our public life, of now literally [putting] money before lives, money before survival. And it leads to a kind of blindness because it's not only cruelty that we're seeing. We're seeing profound ignorance." Sachs also denounced the (then) commander-in-chief, Trump. He was, Sachs said, "the ignoramus-in-chief. He knows nothing, understands nothing. He's a vulgar narcissist." Sachs raised the question of the identity of the experts who guided Congress in spending two trillion dollars. "Our system," he said, "is broken because the greed has supplanted the basic values, and the greed has supplanted people who know what to do."

I presume that Sachs is talking about pervasive corruption. Personally, I feel that the Republicans in Congress do not hide their preferences for the corporate oligarchs. I believe that people, here and abroad, who feel that they have been victimized, may take their anger to the streets. Francesco Rocca, director of the International Federation of Red Cross, reached the same conclusion. He warned of "a social bomb that [could] explode in every moment."[30] He may be right, though the people who suffer the most take time to express themselves. Ignoring the potential social crisis, however, has been typical of oligarchic and plutocratic regimes. We have examples from the agrarian oligarchy of Athens in the early sixth century BCE, to Louis the XVI of France in 1789, to Tzar Nicholas II in 1917 and to the Chinese Revolution of 1949. Unfortunately, it

---

[29]*Democracy Now*, March 24, 2020.

[30]We Need a Public Health New Deal: Neoliberal Austerity & Private Healthcare Worsened U.S. Pandemic, Interview with Gregg Gonsalves and Amy Kapczynski, *Democracy Now*, March 30, 2020.

would seem that billionaires of today did not learn from history. And it appears to me that American billionaires think they can outsource this security threat to the feds. However, things are complicated, especially when the invisible enemy for the most part makes no distinction between rich and poor.

Like Jeffrey Sachs, Gregg Gonsalves,[31] assistant professor of epidemiology of microbial diseases at the School of Public Health at Yale University, was not optimistic about the state that America was in due to the pandemic. He said, "The failure of care in the United States, which is not recapitulated anywhere else in the industrialized world, has led us to a point where [an individual's position in society can determine sickness and access to treatment]." "So," he said, "unless we take care of each other from coast to coast, from north to south, east to west, we [are] going to be [as] vulnerable… as the most vulnerable person in our society. And so, unless undocumented immigrants, unless the incarcerated, unless the homeless are brought into the circle of care, with healthcare universally accessible across the United States, there will always be somebody who [is] going to get sick who could be the spark that sets off the next epidemic." However, his ideas were not highly regarded during the pandemic.

In late-2020, Cate Jenkins, a veteran Environmental Protection Agency (EPA) whistleblower, sent a book-length memorandum to EPA Inspector General Sean O' Donnell urging him to investigate EPA's misleading guidance about the pandemic. She said EPA, the Centers for Disease Control and Prevention (CDC) and the Federal Drug Administration (FDA) misrepresented the time the COVID-19 virus "remains infectious on surfaces," while potent enough to cause disease by eating food containing the virus.[32]

What about Food?

I have been on the periphery of this drama. I clearly identify and sympathize with the concerns of Sachs, Rocca, Gonsalves and Jenkins. Medicine and public health in America are probably no better than agribusiness.

---

[31] *Ibid.*

[32] Cate Jenkins, EPA, CDC and FDA misrepresentations of time the Covid-19 virus remains infectious on surfaces and potential for transmission by ingestion. Memorandum to the EPA Inspector General Sean O' Donnell, Waste Characterization Branch, MRWMD, ORCR, OLEM, EPA, October 26, 2020.

Without citing farming or agribusiness, I think that Sachs, Rocca, and Gonsalves would agree that "health care" is a misnomer. Can anyone equate health care with allowing pesticides in food? As analogies, let us assume that there are some old peoples' homes partially funded by public money. We may then assert that such a function is part of health care. It is only to some degree in my opinion. As for the case of medical doctors, it would probably not be wrong to suggest that some of them may be pushing drugs, not preventing pollution nor preventing diseases. It appears to me that money alone buys medical advice.

In ancient Greek scientific medicine, food was a therapeutic drug.[33] Now the best we may expect is organic food. The Organic Foods Production Act (OFPA) of 1990 forbids the use of chemical pesticides, genetic engineering, sewage sludge, and radiation – in the growing of organic crops.[34] The pandemic was highlighting the advantage of this method of food production. Of all stresses raised by COVID-19, food brought me closer to understanding the violence of the invisible virus. I often walked to the neighborhood grocery store for food during the pandemic. In early 2020, I observed huge crowds forming a line about an hour before the store opened its doors, waiting for their chance to buy food for the family. Once opened, the grocery store managers allowed some 10 people to enter the building every few minutes. Once in the "supermarket," I saw people anxiously and in haste grabbing food, not looking at each other, looking all around, lest they diminished their "social distancing" of 6 feet. I would buy a few organic tomatoes, fruits and vegetables before getting out of the store. But my mind was flooded with images of possibly worse situations, when war, natural disasters, and the looming climate chaos may make it difficult, if not impossible, to find food at all. Food is too important to be at the hands of the very few, as it is now in the United States and many other countries.[35] Scarcity of food or no food is a calamity of unfathomable consequences.

If the virus has a silver lining, let it be this: For as long as he was the President of America, President Biden ought to tackle the food crisis with

---

[33]Lawrence Totelin, When foods became remedies in ancient Greece, *Journal of Ethnopharmacology*, June 5, 2015, 167: 30–37. US National Library of Medicine.

[34]Organic Foods Production Act of 1990. Title XXI – Organic Certification. Sections 2105–2109. Prohibited Crop Production Practices.

[35]Krebs, *The Corporate Reapers*, pp. 440–443; Fowler and Mooney, *Shattering*, pp. 174–200.

equal passion as he would fight the climate change monster. That was how I saw President Biden in 2020. I already said in chapter 3 that President Biden followed former President Obama in food and agriculture. He continued the tenure of Obama's Secretary of Agriculture Tom Vilsack. This was not a wise decision because Vilsack supported agribusiness, not family farmers. Yet Biden was successful in passing bills for funding the rebuilding of America's infrastructure as a step in facing the menace of the climate emergency. Some US$60 billion was dedicated to making agriculture less a source for climate change. Farmers, large and small, received or have an opportunity to receive funds to reduce their carbon emissions and, in general, to support conservation practices like growing cover crops for sequestrating carbon in the soil. Even though the country lost 544,000 small farmers since 1981, and small farmers naturally will find it more difficult to take advantage of available government subsidies, President Biden still deserves credit for rewarding climate-friendly farming.[36]

In the next chapter, I will dissect the anthropogenic climatological and political chaos that are potentially threatening humans and the Earth with dreadful consequences.

---

[36]Lydia DePillis, Can billions in new subsidies keep family farms in business? *New York Times*, May 30, 2024.

6 Climate Chaos

Το Σπίτι ("The House"), painting by Evi Sarantea, 1996. In my personal interpretation, it mirrors the breakdown of the foundations of society and civilization by climate chaos. The surviving house is but the roof, front wall, and door and windows and a couple of children hiding by the door.

The United Nations has been sponsoring the Intergovernmental Panel on Climate Change (IPCC). This scientific organization brings together hundreds of scientists studying the science, origins, and effects of the planet's changing climate. The UN Secretary General, Antonio Guterres, is using the findings of the IPCC studies to urge world leaders to phase out fossil fuels because burning these fuels is causing higher temperatures that imperil humanity and the planet. It is this scientific conclusion, that the burning of petroleum, natural gas, and coal cause climate change, that is making global warming a planetary concern.

An anthropogenic (man-made) warmer natural world is creating new, more violent phenomena that bring chaos to nature and societies. Human industry, the billions of cars, trucks, airplanes, militaries fighting wars, and countless other machines burning petroleum, natural gas, and coal keep pumping solar heat-absorbing gases into the atmosphere. These greenhouse gases affect the constant seasonal temperature of the sun – which is to say, in my opinion, the additional heat contributed by the gases acts like an extra sun. My train of thoughts then assumes that the higher temperatures are disturbing eons-long meteorological and climatological conditions regulating the atmosphere, the waters, and the land of the planet. Early symptoms of this malaise is the acceleration of the destruction of the natural world by outbursts of warmer water in the oceans, for example. In 2014 and 2015, a gigantic ocean heat wave larger than 10 million square kilometers in the West coast of the United States and directly connected to climate change devastated sea lions, salmon, corals, and other marine life.[1] Rising ocean temperatures since 2009 have devastated about 14% of the world's irreplaceable corals.[2,3] Scientists fear heat waves affecting the seas "may become permanent."[4] These heat waves have been afflicting Western Europe and the northern hemisphere. They threaten harvests, water and food, as well as human and environmental

---

[1] Chloe Zilliac, "Return of the Blob: Marine Heat Wave Wreaks Havoc in the Pacific," *Sierra*, October 8, 2019.

[2] Global Coral Reef Monitoring Network, *The Sixth Status of Corals of the World: 2020 Report*, October 5, 2021.

[3] Catrin Einhorn, "Climate Change is Devastating Coral Reefs Worldwide," *New York Times*, October 4, 2021.

[4] Henry Fountain, "Ocean Heat Waves Are Directly Linked to Climate Change," *New York Times*, September 24, 2020.

health. 99% of the world's population breathes unhealthy air.[5] Heat waves, as we said, are harming the oceans. European scientists report that wildlife in the Mediterranean suffers the equivalent of "marine wildfires."[6] They have been warning that "severe heat waves" harm and burn up some sea animals and plants. Sea temperature in the Mediterranean in the Summer of 2022 was "largely out of sight and out of mind."[7]

Threatening Primordial Plankton and the Melting of Life

Even the very foundation of marine life, plankton, are at risk. Petroleum companies discover oil and gas in the oceans by causing local earthquakes underwater. They employ seismic airguns with such exploding power that they devastate the zooplankton and krill, both food for fish and marine mammals like whales. "[S]cientists found that the blasts from a single seismic airgun caused a statistically significant decrease in zooplankton 24 hours after exposure. Abundance fell by at least 50% in more than half (58%) of the species observed. The scientists also found two to three times more dead zooplankton following airgun exposure compared to controls and, shockingly, krill larvae were completely wiped out," wrote Francine Kershaw of the environmental organization Natural Resources Defense Council.[8]

European scientists studying plankton are extremely concerned with their dramatic decline in some regions of the Atlantic. Scientists of Global Oceanic Environmental Survey (GOES) Foundation theorize that the causes of plankton decline include increasing carbon dioxide in the oceans, and water pollution by synthetic chemicals, plastics, cosmetics, sunscreen, drugs and agricultural pesticides and fertilizers. It's possible that "the horrendous [plankton] falloff is a result of migration north

---

[5] UN General Assembly declares access to clean and healthy environment a universal right, *UN News*, July 28, 2022.

[6] Jon Henley, Mediterranean ecosystem suffering wildfire as temperatures peak, *The Guardian*, July 29, 2022.

[7] Ciaran Giles and Ilan Ben Zion, Scientists warn of massive mortality of marine life as Mediterranean Sea heats up, *Los Angeles Times*, August 17, 2022.

[8] Francine Kershaw, New science: Seismic blasting devastates Ocean's Zooplankton, *NRDC*, June 22, 2017.

because of too warm waters or decimation by toxins such as plastics, chemicals, and farm pesticides/fertilizers. However, it's likely a combination of factors, which essentially doubles the trouble."[9]

Other destructive effects of the anthropogenic climate change include heat waves, ice melting, cyclones, storms, fires, and floods. Antonio Guterres of the United Nations said the impact of humans was just like that of the meteor that wiped out the dinosaurs some 70 million years ago. "In the case of climate," he said, "we are not the dinosaurs. We are the meteor. We are not only in danger. We are the danger."[10]

My understanding of the danger that humans pose is our industries and our use of fossil fuel that melts the ice bergs, breaking up the ice in Antarctica and Alaska. These unfortunate events reverberate around the world. Ice melting on land and seas on a gigantic scale is primordial in my viewpoint. Equally concerning is the thawing of permafrost in Siberia and Alaska (Fig. 1). Unmeasured but suspected vast amounts of frozen carbon are seeping from the depths of the frozen land to the atmosphere. "The frozen ground," CNN, using NASA data, reported in 2022, "holds an estimated 1700 billion metric tons of carbon – roughly 51 times the amount of carbon the world released as fossil fuel emissions in 2019."[11]

I heard more dreadful stories in China. In my November 2019 visit to that country, I met a former German politician, Hans-Josef Fell. In our conversation, he spoke about climate change as well. He worried about existential threats if the world leaders continue to delay cleaner energy. Fell echoed the UN climate experts, though their language is more restrained and diplomatic. They warned in 2019 that humanity has about a decade to prevent "irreversible damage" to the natural and social world from the anthropogenic climate change.[12] In essence: basic questions for survival must be resolved in the next ten years. On November 5, 2019, 11,000 scientists from 150 countries issued a warning to the leaders of

---

[9] Robert Hunziker, Breakdown of the marine food web, *Counterpunch*, July 29, 2022; See a summary of the arguments on the decline of Plankton in Beyond Pesticides Daily News Blog, July 22, 2022.

[10] United Nations, Secretary General Antonio Guterres, A moment of truth, June 5, 2024.

[11] Katie Hunt, Belching lakes, mystery craters, zombie fires: How the climate crisis is transforming the Arctic permafrost, *CNN*, November 12, 2022.

[12] UN General Assembly, Seventy-Third Session, Only 11 years left to prevent irreversible damage from climate change, March 28, 2019.

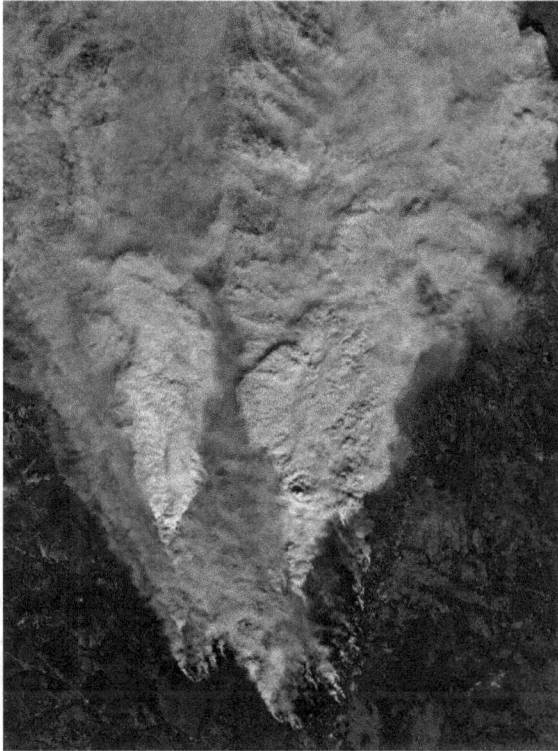

Fig. 1.   Permafrost forest on fire. Abyysky district, Yakutia, Siberian Arctic, northern Russia. Photo: July 10, 2024. Courtesy of Earth Observatory, NASA.

the world.[13] "Despite 40 years of global climate negotiations," they said, "we have largely failed to address this predicament... The climate crisis has arrived and is accelerating faster than most scientists expected... It is more severe than anticipated, threatening natural ecosystems and the fate of humanity... Especially worrisome are potential irreversible climate tipping points and nature's reinforcing feedbacks (atmospheric, marine, and terrestrial) that could lead to a catastrophic "hothouse Earth," well beyond the control of humans... These climate chain reactions could cause significant disruptions to ecosystems, society, and economies, potentially making large areas of the Earth uninhabitable."

---

[13] William J. Ripple *et al.,* World scientists' warning of a climate emergency, *BioScience*, November 5, 2019.

China and America

President Xi Jinping of China would do well to heed the advice of these scientists and dramatically cut China's gigantic carbon emissions. I have been arguing in this book that banning animal farms and the hunting of wildlife would help human efforts in fighting climate change, as well as lessen diseases and pandemics. Despite the rising Cold War tensions between China and the United States,[14] the two countries could potentially work together to systematically reform our obsolete traditions loaded with inequality, bad science, endless pollution, population increase, and nuclear bombs. The leaders of both countries must fight climate change together. Think big and act to save the world.

During the annual UN General Assembly meeting on September 22, 2020, Xi Jinping pledged that China would fight climate change.[15] He spoke boldly of the good intentions of China to make Mother Earth a better home for all and, simultaneously, convert China from the world's number one contributor to global warming to one with zero carbon emissions before 2060. "COVID-19 reminds us that humankind should… make Mother Earth a better place for all," said Xi …. The Paris Agreement on climate change [signed by 195 nations, April 22, 2016] charts the course for the world to transition to green and low-carbon development. It outlines the minimum steps to be taken to protect the Earth, our shared homeland, and all countries must take decisive steps to honor this Agreement. China will scale up its Intended Nationally Determined Contributions by adopting more vigorous policies and measures. We aim to have CO_2 emissions peak before 2030 and achieve carbon neutrality before 2060…. China is the largest developing country in the world, a country that is committed to peaceful, open, cooperative and common development. We will never seek hegemony, expansion, or sphere of influence. We have no intention to fight either a Cold War or a hot war with any country. We will continue to narrow differences and resolve disputes with others through dialogue and negotiation.

In my opinion, pledging a carbon neutral China before 2060 muddies the waters of green development and protecting Mother Earth.

---

[14] Alice Su, China orders closure of a U.S. consulate, *Los Angeles Times*, July 25, 2020.

[15] Xi Jinping, Statement at the General Debate of the 75th Session of the UN General Assembly (Ministry of Foreign Affairs of the People's Republic of China, September 22, 2020).

It evaporates any notion of ecological civilization. The bold promise of peace and science-based decisions do not square with the reality of climate danger threatening China and the world. The target of a carbon-neutral China in 2060 unfortunately does not meet the deadline posted by scientists who said that decisive actions on climate change is necessary before 2030 if we wish to save the planet from irreversible destruction. At the very least, greenhouse gas emissions must be reduced by 50% by 2030, so that global temperature does not exceed 1.5°C (2.7°F) above the temperature of the pre-industrialization era. Nonetheless, Xi Jinping's announcement is still significant because anything that China does to reduce its greenhouse emissions immediately translates to a slower global warming.[16] In September 2021, Xi Jinping promised China would no longer finance and build coal-fired plants abroad.[17] It would help if America's Cold War against China changed to cooperation instead. A group of American environmental organizations sent a letter to President Joe Biden and members of Congress on July 7, 2021, in which they pleaded for a common American and Chinese front for fighting climate change. They said:

> "[W]e are deeply troubled by the growing Cold War mentality driving the United States' approach to China – an antagonistic posture that risks undermining much-needed climate cooperation… Amid a climate emergency that is wreaking havoc on communities across the globe, the path to a livable future demands new internationalism rooted in global cooperation, resource sharing, and solidarity. Nothing less than the future of our planet depends on ending the new Cold War between the United States and China."[18]

No doubt America and China must work together. Perhaps, the United States could stop arming Taiwan, and China could stop producing more nuclear bombs, hypersonic missiles, and "exotic delivery technologies"

---

[16] Somini Sengupta, China, in pointed message to U.S., tightens climate targets, *New York Times*, September 22, 2020.

[17] Azi Paybarah, China says it won't build new coal plants. What does that mean? *New York Times*, September 22, 2021.

[18] Letter to President Biden and Members of US Congress, Friends of the Earth, July 7, 2021.

for nuclear weapons.[19] The growing number of Chinese warheads is "worrisome." China's buildup of nuclear weapons certainly raises the stakes.[20] Unchecked Chinese strategic competition with America, and an unrestrained United States Cold War against China, all but doom the world as we know it. In other words, China, the United States, and Russia must find a way to work together for two very important reasons: dismantling the world's monstrous nuclear weapons and joining hands in transitioning the world from fossil fuels to cleaner energy – soon. Fortunately, during the 26th UN climate conference of the parties (COP 26) in Glasgow, Scotland, United Kingdom, from October 30 to November 12, 2021, representatives of China and the United States promised to accelerate cooperation in diminishing methane and coal burning emissions.[21] The UN Secretary General, Antonio Guterres, welcomed environmental cooperation between China and the United States over climate as an an important "step in the right direction."[22] For the first time in global climate meetings, negotiations took place among 200 countries on phasing out coal.

The Oil Flowing Deal

But these negotiations, like other commitments of individual nations, are barely tied to the obligations of treaties and are not enforceable. The Earth Summit in Rio de Janeiro in Brazil in 1992 did not put any restrictions on greenhouse gas emissions. The Kyoto Protocol of 1997 in Kyoto, Japan and the Paris Agreement of 2015 include legal obligation concerning the use of fossil fuels and the reduction of greenhouse gas emissions from the burning of those fuels. The United States twice abandoned the international climate accords, under President George W. Bush and President Donald Trump. Bush opposed the Kyoto Protocol[23] and Trump the

---

[19] Tong Zhao, China's silence on nuclear arms buildup fuels speculation on motives, *Bulletin of the Atomic Scientists*, November 12, 2021.

[20] Alice Su, China's nuclear weapons buildup raises the stakes, *Los Angeles Times*, November 17, 2021.

[21] UN Climate Action: COP 26: Together for our planet.

[22] UN News, UN chief welcomes China-US pledge to cooperate on climate action, November 10, 2021.

[23] The White House, Office of the Press Secretary, Text of a Letter from the President [George W. Bush] to Senators [Opposing the Kyoto Protocol], March 13, 2001.

Paris Agreement.[24] No wonder Jennifer Morgan, director of Greenpeace International, said the promises at Glasgow were not about to solve the climate emergency or to give confidence to our children that the future was secure.[25]

There is also the history of war connected to petroleum. In February 1945, President Roosevelt met with King Abdulaziz bin Saud of Saudi Arabia where they agreed to exchange Saudi oil for American military protection of Saudi Arabia. Most of the American Presidents supported the kingdom of Saudi Arabia during their times of need, with the exception of President Donald Trump during his first administration, and the less than enthusiastic response from President Biden during the recent skirmishes in the Middle East due to political differences.[26] The reason is the vast quantities of petroleum in the sands of Saudi Arabia. America's colossal appetite and use of petroleum depends on easy access to Saudi oil. It is this military connection to petroleum that is a tremendous obstacle in moving the United States rapidly out of the fossil fuel era.

The Year 2030

Despite this and other inconvenient truths tying America and other industrialized countries to fossil fuels, the entire world has to act fast before 2030. In their August 9, 2021 report, the international scientists of the UN Intergovernmental Panel on Climate Change warned world leaders that the temperature of the planet in 2021 was already 1.09°C above pre-industrial levels. In other words, global temperature rose a little more than 1°C since 1850–1900. UN scientists predicted that if world leaders continue to do nothing, it will bring global temperatures up 1.5°C by 2030.[27] That is the main reason why the year 2030 has become so critical in defining the fate of the planet and civilization.

I have been highlighting in this book that rising global temperature is responsible for climate chaos: more heat waves, intense rains and

---

[24]US Department of State, On the U.S. Withdrawal from the Paris Agreement Press Statement, November 4, 2019.

[25]Seth Borestein, Ellen Knickmeyer and Frank Jordans, China, U.S. pledge to increase cooperation at U.N. climate talks, *Los Angeles Times*, November 10, 2021.

[26]Jenny Spalding, The deal that keeps the oil flowing, *Epicenter*, June 1, 2023.

[27]Intergovernmental Panel on Climate Change, *Climate Change 2021: Physical Science Basis*, August 9, 2021.

flooding, droughts, sea level rise, permafrost thawing, melting of glaciers and ice sheets, loss of Summer Arctic Sea ice, marine heatwaves, ocean acidification, and sea dead zones. Even snow is becoming rare. There was plenty of snow in late 2024 – early 25. Yet the scarcity of snoe in the past few years is of great concern, especially for the water available in the American West.[28] Having little or no water will affect food crops, other plants, the land, wildlife, wildfires, and human beings. We have a few years left during which we can, at least, reverse and limit the drastic effects of climate change.[29] Scientists have been raising the alarm that we have to prevent global temperature rising above 1.5°C in order to prevent a climate catastrophe. To Larry Schweiger, environmentalist since the 1960s, senior executive of the National Wildlife Federation, and an insider to the misleading policies of the fossil fuel industry, the task of putting a brake to the forthcoming climate catastrophe is going to be painful and probably politically unpalatable. In 2019, he urged us to declare war on climate change. WWII-like footing, he says, is necessary "to avoid deadly droughts, mass starvation, dread diseases, fierce coastal storms, and widespread floods, forest fire destruction, forced migrations, and border wars."[30]

On February 23, 2021, John Kerry, President Joe Biden's climate envoy, told the UN Security Council that the time had come to address the climate monster. He expressed regret that President Trump had misled America and the world about global warming. But continuing Trump's irresponsible policy, ignoring climate change would be the equivalent to the UN members supporting "a mutual suicide pact."[31] The UN Chief, Guterres, has been talking about a mutual suicide pact and climate high-ways of hell. On June 5, 2024, he said: "We are playing Russian roulette with our planet. We need an exit ramp off the highway to climate hell, and the truth is we have control of the wheel."[32] Yet politicians remain silent, seemingly obedient to the dictates of the fossil fuel industry.

[28] Erica R. Siirila *et al.*, A low-to-no snow future and its impacts on water resources in the Western United States, *Nature Reviews Earth and Environment*, 2, pp. 800–819, October 26, 2021.

[29] Bill McKibben, A very hot year, *New Yorker*, March 12, 2020.

[30] Larry Schweiger, *Climate Crisis and Corrupt Politics: Overcoming the Powerful Forces that Threaten our Future* (Irvine, Boca Raton: Universal Publishers, 2019) 219.

[31] Somini Sengupta, Biden's climate envoy, at U.N., likens global inaction to a "suicide pact," *New York Times*, February 23, 2021.

[32] Antonio Guterres, A Moment of Truth, *UN News*, June 5, 2024.

In late 2022, two American scholars, Wes Jackson, and Robert Jensen, compared the risks of climate change to an apocalypse devoid of its theological meaning. Jackson is an agricultural expert, and Jensen is a professor of journalism. They are not accustomed to exaggeration. Jackson, in particular, has spent his life trying to stop the goliath of agribusiness from devouring rural America. He has been developing perennial seeds and crops. His Land Institute in Salinas, Kansas, has been a laboratory of science and revolution. The perennial seeds he is developing are designed to bring us food without the use of the pesticides and biocides of agribusiness.

Wes Jackson and Robert Jensen speak of inconvenient truths like those promoted by Guterres. They say that we certainly knew for a very long time that our world is "on the brink of an apocalypse" and yet we have done very little to diminish the extreme danger of our anthropogenic crises – in unsettling the natural world, including the climate of the planet. Governments and philanthropic foundations and universities, Jackson and Jensen say, fund only projects that agree with their outdated polices. Governments refuse to face the truth that continuing today's levels of the burning of fossil fuels is the equivalent of setting Earth on fire. Jackson and Jensen reject the rhetoric of the powerful and their myth of economic growth and technological progress. "The end result," they say, "will be apocalyptic; the only question now is how bad it will be."[33]

I agree. In fact, four of my books speak about those bad effects and the lost opportunity of protecting our planet from easy and feel-good "development" projects, agribusiness, pesticides salesmen and a myriad of other hucksters of profit at any price. I think that the social, political and ecological situations in America and the rest of the world hinge on our abilities to change. Perhaps the idea of a climate apocalypse may trigger deeper thoughts on how we can avoid such a catastrophe. We still have a chance to reverse the extreme consequences of climate chaos, if the people of the world begin to listen and act upon the warnings given by scientists. My suggestion is to start this struggle by banning fossil fuels and replacing them with renewable and non-polluting energy. Leaving fossil fuels in the ground would be a boon to public health. Stopping the burning of fossil fuels is following science, and the beginning of taking

---

[33] Wes Jackson and Robert Jensen, *An Inconvenient Apocalypse: Environmental Collapse, Climate Crisis, and the Fate of Humanity* (Notre Dame, IN: University of Notre Dame Press, 2022), pp. 127–139, back cover.

nature seriously. A drastic reduction in greenhouse gases emissions into the atmosphere would put a brake on global warming. Healthy alternatives exist as we can get the energy we need from solar power and other non-polluting sources like wind, geothermal energy and water. I am not predicting the future with empty words when I say that we need to act fast to save ourselves and the planet from our petroleum nightmares. I am a historian who has been observing American relations with the natural world for decades, I have studied the dark history of human exploitation of the Earth, and am a witness to China wrestling with its history and corrupt choices of our times.

Nemesis

The natural world is not vengeful, and neither were the Greek gods. Greek culture was infused with the idea of perfection in moderation. Monopoly was regarded as anti-democracy in society and the Cosmos. This moral order appeared to come apart with the upsurge of monotheistic religions whereby power became a tool for the leaders.[34] In my view, the arbitrary power of religions, monarchies and oligarchies likely influenced the rise of the monopoly of oil and gas as well as coal companies. The financial returns of the two industries are expected to be huge since fossil fuels power armies, industries and societies. In the absence of moral and ecological global leadership, fossil fuel companies may stay in business, no matter the effects of making the planet hostile to life.

Injustice envelops America and the world from the overwhelming power of fossil fuel corporations and their political counterparts who are destabilizing the planet with their temperature-rising products. The Earth took billions of years of evolution to bring to life millions of beautiful species of plants and animals, including humans. This injustice takes substance and voice from science, with the scientists warning us that the extraction and burning of fossil fuels is setting the Earth on fire, threatening the planet; killing the primordial plankton at the floor of the oceans, burning forests, harming all life in the oceans, other ecosystems, and bringing fierce destruction to human communities all over the world.

---

[34] Evaggelos Vallianatos, *The Passion of the Greeks: Christianity and the Rape of the Hellenes* (Harwich Port, Cape Cod: Clock and Rose Press, 2006).

All these issues bring to mind Nemesis, the Greek goddess whose name translates into "retribution and wrath."

I tasted the human-made Nemesis in 1989–1990. I wrote my first article about petroleum companies causing climate change. The article[35] got me into trouble with senior officials at the US Environmental Protection Agency. This article appeared in the *Chicago Tribune* on October 10, 1989. Six months later, on Earth Day 1990, the *Wall Street Journal* reprinted a couple of paragraphs from my article. These paragraphs zeroed in on the industry harming ecosystems and people from burning petroleum. That sufficed to trigger reactions from Capitol Hill and a concerted effort within the EPA to fire me. However, to his credit, the administrator of EPA, William Reilly, blocked the action of the EPA petroleum enthusiasts – and saved my job. From that moment on, my focus on the climate crisis increased.

Why is the US government and Americans tolerating an increasing danger to their lives and the health of the natural world from the burning of petroleum, natural gas, and coal? The science is straightforward. The burning of fossil fuels is increasing the temperature of the planet, which harms people and the environment. So, the solution is obvious: Stop burning fossil fuels. But, as we know it, the solution is not easy to execute because the entire economy and society, as well as the world, are dependent on burning the very substances that are causing local and global harm, disease, and destruction. The US EPA reported that climate change brings diseases like "respiratory and heart diseases, pest-related diseases like Lyme disease and West Nile Virus, water- and food-related illnesses, as well as injuries and deaths. Climate change has also been linked to the increase in violent crime and overall poor mental health."[36]

Contemporary society did so much damage to wildlife and the natural world that man has become the overwhelming species and top predator. Scientists are expressing this dominance of man by describing our epoch as the Anthropocene – from "*anthropos*" which means "man" in Greek. Men (primarily of the industrialized West) turned the science and technology they borrowed and advanced from the Greeks into weapons for the conquest of the world – including land, the seas, and the sky. That "conquest" came through the expansion of Europe to the tropics and

---

[35] Evaggelos Vallianatos, Lower the Earth-threatening heat, *Chicago Tribune*, October 10, 1989.
[36] US Environmental Protection Agency, Climate change and human health, June 4, 2024.

the Americas. The new European rulers started deforesting and industrial-izing the lands they occupied. This transformation, at home and in the colonies, brought immense and unforeseen calamities: nearly wiping out the rich variety of life and filling the lands with poisons and the seas with plastics, oil, and deleterious pollution.[37] You cannot go on killing and forcing to extinction wild animals and plants, including insects, birds, fish, and other countless forms of life on land, rivers, lakes and the seas without a response from the natural world.

In addition, there is an overwhelming number of more than 8 billion humans populating the Earth. The physician Warren Hern is convinced humans are becoming a cancer and a disease on the planet.[38] According to the British charity, Population Matters,[39] some 10,000 years ago, humans made up 1% of the animal population. Wild animals were the overwhelm-ing majority: 99%. In 2011, humans were 32% and wild animals 1% of the animal population. About 67% of non-wild animals were domesticated and bred for human food. In my point of view, the missing wild animals, even the tiniest, were the species that kept the Earth beautiful, fruitful, habitable, and alive.

This unfortunate reversal in wild animal population, the ceaseless economic demand for growth, the perpetual increase of human population and the inevitable huge consumption of resources broke the camel's back. Billions of humans are occupying all the ecological zones of the planet where they have been using forests, minerals, animals, and fish for liveli-hoods and for making money. Such a development turned things upside down in the natural world. The dominant role of humans regrettably set in motion a dramatic decline among all species of plants and animals. Scientists speak of the annihilation of vertebrate animals in the sixth mass extinction.[40]

---

[37]Katie Pavid, What is the Anthropocene and why does it matter? Natural History Museum.

[38]Clay Bonnyman Evans, Doctor's diagnosis for the Earth: A terminal human malignancy, *Colorado Arts and Sciences Magazine*, November 8, 2022.

[39]Population Matters, Nature's decline deepens: Insights from the living planet report 2024.

[40]Gerardo Caballos *et al.*, Biological annihilation via the ongoing sixth mass extinction signaled by vertebrate population losses and declines, *Proceedings of the National Academy of Sciences*, July 10, 2017.

Climate change was inevitable. It can even be said that power and easy profits corrupted the humans who discovered fossil fuels. They found out that petroleum, natural gas, and coal profits earned them princely positions in the hierarchy of running states, economies, and the world. In my opinion, the fossil fuels owners then used their positions unfairly to influence economies, technologies, politicians, and governments to power industry, transportation, industrialized agriculture, the heating of homes and power countless other machines using their products. Cars and trucks powered by petroleum became the embodiment of individualism and capitalist profit. City planning accommodated the imperial car. Serving the public good and the health of the natural world by public transport became secondary. A century later, the supremacy of fossil fuels proved catastrophic. Fossil fuels became a planetary monoculture and global warming sprang from that vast disequilibrium. Even the most ancient human activity of raising food became a subsidiary of petroleum – at least in the so-called developed world which comprises Europe, North America, Australia, China, and Japan. On September 22, 2024, the UN Secretary General said, "The climate crisis is destroying lives, devastating communities and ravaging economies. We all know the solution – just phase out of fossil fuels – and yet, emissions are still rising."[41]

There's little doubt in my mind that Guterres is right. We should act now (in the next few remaining years of the 2020s) to make the fundamental changes necessary in slowing down the awakened climate monster. Start with fundamentals. Food and agriculture. Reforming industrialized farming to traditional agriculture without pesticides and synthetic fertilizers would boost biological diversity and pollinators like honeybees. Traditional agriculture is "highly sustainable and productive." It is also rich in biodiversity and preserves biodiversity and natural resources.[42,43] The transformation of industrialized farming to a more nature and human health friendly agriculture would help human health. And, certainly, synthetic chemicals-free farming and cropland would decrease the impact of climate change. It will make it less destructive because, like forests,

---

[41] Secretary-General remarks at the Opening Segment of the Future Plenary, *UN News*, September 22, 2024.

[42] Miguel A. Altieri, Applying agroecology to enhance the productivity of peasant farming systems in Latin America, *Environment, Development and Sustainability*, September 1999.

[43] M. A. Altiery, Traditional Agriculture, *Encyclopedia of Biodiversity*, 2001.

traditional farming captures carbon dioxide from the atmosphere and puts it in the soil. Humans, all other animals, plants or crops, as well as fruits and vegetables need healthy land and food, certainly uncontaminated by toxins.

Keeping things as they are is not an option. Each second, humans add so much carbon and methane to the atmosphere that the heat trapped by those emissions is the equivalent of blowing up four Hiroshima atomic bombs.[44] In May 2020, the Mauna Loa National Laboratory in Hawaii measured 417 parts per million (ppm) of carbon dioxide, CO_2, in the air. This was unprecedented in climate history. For countless millennia, carbon dioxide stayed at around 300 ppm in the atmosphere. In 2014, the carbon monthly values at Mauna Loa reached the level of 400 ppm. But the 417 ppm of carbon in the atmosphere broke the record for several million years.[45] Science says that this high value of carbon pollution has been the result of the burning of fossil fuels in our various industries and societies – adding about 35 billion metric tons of carbon to the atmosphere every year. In 2019, the total amount of carbon dioxide that reached the atmosphere was 43.1 billion metric tons.[46]

Business as usual, fueled by the continuing burning of petroleum, natural gas, and coal, has been responsible for increasing the amounts of carbon dioxide in the atmosphere. "Business as usual" means businesses operating as if there was no evidence of climate change; namely the owners of business convincing themselves that their factories, cars, trucks, airplanes, and boats make no impact on the climate, or business owners rejecting science and concluding that climate change is a hoax. In either case, business as usual ignores science and is therefore anachronistic, irresponsible, and dangerous. Business owners must understand they are primarily responsible for the climate emergency threatening America and the rest of the world. CO_2 is a heat-trapping gas formed during the extraction and burning of petroleum, natural gas, and coal as well as from wildfires and the eruption of volcanoes. Carbon dioxide and other greenhouse gases warm the planet.

In the case of America's gigantic emissions of greenhouse gases, the unspoken elephants in the room are the Pentagon and agriculture. Their

---

[44] Skeptical Science is a joint effort for recording talks. John Cook gave a talk about Climate Change for Skeptical Science, November 25, 2013.

[45] Robert Monroe, *Climate Change*, UC San Diego Scripps Institution of Oceanography, June 4, 2020.

[46] Chelsea Harvey and Nathannial Gronewold, Greenhouse gas emissions to set new record this year, but rate of growth shrinks, *Science*, December 4, 2019.

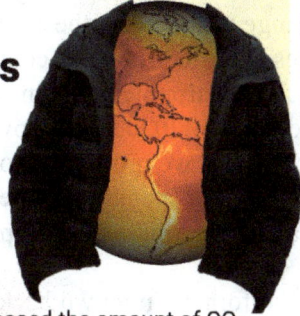

Carbon dioxide traps heat, like a puffy coat

Humans have increased the amount of CO₂ in our atmosphere by 50% since 1750.

We've made Earth's insulating "puffy coat" thicker than it's been in millions of years. As a result, our climate is warming rapidly.

CLIMATE.NASA.GOV

Fig. 2. Carbon dioxide trapping heat. Credit: NASA/JPL-Caltech, October 26, 2022; David Gelles, Three Greenhouse Gases, Three All-Time Highs, *New York Times*, April 9, 2024.

carbon footprint is very large. From 2001 to 2017, the US military greenhouse gas emissions were equivalent to 1.2 billion metric tons. The Department of Defense (DOD) is the world's largest institutional user of petroleum and "the single largest producer of greenhouse gases in the world."[47] In 2012, one-third of global greenhouse gas emissions came from agriculture.[48] In the United States, during Trump's 2017–2021 term as President, we did not know the exact contribution of agriculture to making our planet warmer. But my belief is that agribusiness (especially the growing of conventional crops and animal farms) was responsible for more than one-third of all of US's greenhouse gas emissions. The British naturalist David Attenborough[49] found it hard to believe that

---

[47] Neta C. Crawford, Pentagon fuel use, climate change, and the costs of war, *International and Public Affairs*, Watson Institute, Brown University, June 12, 2019.

[48] Natasha Gilbert, One-third of our greenhouse gas emissions come from agriculture, *Nature*, October 31, 2012.

[49] Louise Boyle, The present administration is disastrous. David Attenborough condemns Trump's impact on the natural world, *Independent*, September 28, 2020.

Trump's America was in an appalling climate stupor. The terrible danger of climate change was all over the country, Attenborough said. Cyclones and hurricanes were tearing the country apart. The ferocity and frequency of such catastrophic phenomena were unprecedented as evident from the wild fires in 2020.[50] Moreover, a "warming climate will alter everything from how we grow food to where people can plausibly live. Ultimately, millions of people will be displaced by flooding, fires and scorching heat, a resorting of the map not seen since the Dust Bowl of the 1930s."[51]

In 2020, California – my own backyard – suffered from a heat wave. I have lived in Claremont, California, since 2008, and never felt as much heat as I did on Saturday, September 5, 2020. That was a very hot day. The temperature rose to about 120°F (~49°C). I usually read and write under the portico in my backyard. But that Saturday and the next day, Sunday, September 6, 2020, I did not. The heat in the atmosphere coupled with the heat from the bright and life-giving Sun formed a hot, aggressive wave that enveloped me like I was walking towards a huge fire oven. I felt warmth all over my body, and the temperature was still rising as the day progressed; my brain then was telling me to seek shelter in my home. My thoughts immediately rushed to the incredible discomfort of the two doves that lived in a corner of the roof of my portico. They were trying to hatch an egg for another chick. I saw them changing guard over the egg, staying motionless for hours under this burning heat. At that point, I felt helpless and very angry – for them and us. My thoughts also brought my own plight to the fore. Even when the temperature dropped lower to the 90s later in the day, the discomfort continued for hours and, sometimes, the entire day and night.

The US House Committee on Oversight and Reform spent 2021–2022 investigating America's oil companies' profiteering and greenwashing connections to climate change. Chairwoman Carolyn B. Maloney and Subcommittee Chairman on the Environment, Ro Khanna, accused these companies of failing to take responsibility for the harm their

---

[50] PBS Newshour, September 10, 2020.

[51] Lucas Waldren and Abrahm Lustgarten, Climate change will make parts of the U.S. uninhabitable. Americans are still moving there, *ProPublica*, November 10, 2020.

petroleum has already done in causing climate chaos. Chairwoman Maloney said:

> *Even though Big Oil CEOs admitted to my Committee that their products are causing a climate emergency, today's documents reveal that the industry has no real plans to clean up its act and is barreling ahead with plans to pump more dirty fuels for decades to come. Today's new evidence makes clear that these companies know their climate pledges are inadequate but are prioritizing Big Oil's record profits over the human costs of climate change. It's time for the fossil fuel industry to stop lying to the American people and finally take serious steps to reduce emissions and address the global climate crisis they helped create.*[52]

When *New York Times* opinion writer David Wallace-Wells interviewed Rep. Ro Khanna of California, Khanna said: "[America's oil companies exhibit] this pervasive bunker mentality – hunker down and fight the external world that cares about climate… I was also struck by the bullying, the vitriol against climate activists and climate reporters… Most people in this country don't know that Big Oil has lied for decades about climate change. That's just not on their radar. Now these companies are positioning themselves as clean companies, they're doing it in a very shrewd way, saying: 'We're going to do things about our operations.' But the documents show that they basically have a plan that this is going to give them, as they put it, a license to operate… But what I would say about Big Oil is that they have not made in any way substantively a transition in diversifying. But at the same time, they're telling the public that they are. That's the most jarring part of this – they want to be seen as good guys."[53]

Imagine, polluters undermining civilization and harming the entire planet wishing others to see them as good guys! But let's take this power another step further. Seeing in your mind's eye the stubbornness of the fossil fuel industry and its vigorous supporters in the Republican and

---

[52] House Committee on Oversight and Reform, Internal documents reveal the industry is making long-term fossil fuel investments as they resist and block regulation, December 9, 2022.

[53] David Wallace-Wells, Big oil companies are bullies who want to be seen as good guys, *New York Times*, December 15, 2022.

Democratic governments may tempt you to dismiss all the talk about climate change and keep your air conditioner on. Cease thinking. Do nothing. This is no different than the edict of Augustine of Hippo (354–430 A.D.), who said that curiosity is a disease.[54] Under such circumstances, it's tempting to keep the status quo. However, curiosity is not a disease. It is rather the motor power of science and truth. It is curiosity that helped us discover the mistake of burning fossil fuels.

During very hot days, I sought relief from the sweltering temperatures under my trees. I kept admiring them for their survival and the fruits they kept giving me year after year. But in the Summer of 2022, most of them gave me little, if any fruit, save for the abundance of my lemon tree and an orange tree in my backyard. My fig tree in the front yard lost its leaves by mid-August. These visible reminders made me wonder why we are continuing to fuel the vast, complicated and dangerous climatological engine warming the planet? Is the lives of most humans so worthless that we have no qualms trading them away for the power of petroleum companies? This prospect should have been unacceptable.

The Inferno of California and the West Coast

A few weeks in August and September 2020 became a climate catastrophe for the West Coast. A shattering heat wave and fires embraced the region.[55] Joseph Serna, a reporter from *Los Angeles Times* said in an article that an "epic firestorm" wiped out several towns and killed dozens of people in California, Washington and Oregon.[56] These fires were excessively violent with long-lasting ecological consequences. Ryan Bauer, a Californian fire expert, said: "I think of [any of these destructive fires] more as an avalanche of fire... It's not just rolling down the hill, it's compounding itself and making itself worse and more powerful as it goes, starting more fires and then this cloud of embers ahead of it lighting

[54] *The Confessions of St. Augustine* 10.35, tr., Rex Warner (New York: The New American Library, 1963).

[55] Abrahm Lustgarten, Climate change will force a new American Migration, *Propublica*, September 15, 2020.

[56] Joseph Serna, The state's 2020 wildfires will leave many ecosystems altered for centuries, *Los Angeles Times*, December 24, 2020.

additional fires."[57] The climate firestorm filled the air with pollution, the worst ever.[58] A friend from Eugene, Oregon, Linda Reed Haase, sent me an email on September 16, 2020, in which she said: "The air quality [in Eugene] is sooo bad!!! I'm truly afraid of what we are going to see when visibility returns. Of course, where I live is perhaps 10 miles from the closest area that was being evacuated, so fire damage is not visible from here, but we're going to see little towns completely destroyed, burned out landscape and old growth stands just gone. That is horrific and tragic, but then add to it the awful human tragedy of peoples' homes and all possessions lost, displacement and nothing to go back to. It is heartbreaking."

Heartbreaking, indeed. West Coast cities like Eugene, Portland, Seattle, Los Angeles and San Francisco were overwhelmed by very bad air pollution and smoke – gigantic clouds of smoke. The 8648 wildfires in California in 2020 burned about five million acres of forest and urban land.[59] California Governor Gavin Newsom said the fires cleared his mind about climate change. He said there was no question about what fueled the fires: "The debate is over around climate change," he said. "Just come to the state of California. Observe it with your own eyes. It's not an intellectual debate. It's not even debatable any longer, what we are experiencing – the extreme droughts, the extreme atmospheric rivers, the extreme heat. Just think. In the last few weeks [of August and September 2020] alone, we've experienced the hottest August in California history. We had 14,000 dry lightning strikes over a three-day period. We're experiencing temperatures, world record-breaking temperatures, in the state of California, 130 degrees [Fahrenheit]... California, folks, is America fast forward. What we're experiencing right here is coming to a community all across the United States of America unless we get our act together on climate change, unless we disabuse ourselves of all the BS that's being spewed by a very small group of people."[60]

Despite the brave words, Newsom, in my opinion, was not serious about the inferno burning California. Since 2020, he failed to change his

---

[57] Joseph Serna, The state's 2020 wildfires will leave many ecosystems altered for centuries, *Los Angeles Times*, December 24, 2020.

[58] Rong-Gong Lin II and Joseph Serna, Extreme weather stoked wildfires, *Los Angeles Times*, September 14, 2020.

[59] State of California, 2020 Incident Archive.

[60] Democracy Now, September 14, 2020.

Fig. 3.   Smoke from a nearby forest fire envelops the San Francesco Golden Gate Bridge with near invisibility, 2018. Photo by Evaggelos Vallianatos.

policies regarding forest protection, agriculture, transportation, solar power and conventional electricity production. His policies still favored business and conventional energy in my view. Instead, forest fires became almost "normal." Television and newspaper reporters speak of fire seasons. They are right in their predictions. Fires do incinerate public forests and, sometimes, towns. This annual prediction of forest fires, however, leaves out the ecocidal effects of fires. Let us not forget that smoke and flames mix with the atmosphere, including the air people breathe too. It would appear that smoke briefly makes important decisions about climate change invisible (Fig. 3).

"[T]he dark orange panoramas of the 2020 [fire] season [became] the ready picture of the burning West, the color almost matching the towers of the Golden Gate Bridge and the shades of flickering fires in the distance but held aloft above America's city of the future, as though the sky itself were in flames and ready to be breathed – indeed, as though there were nothing else to breathe," wrote David Wallace-Wells of the *New York Times*.[61]

Despite this threat, neither California nor the United States acted as they should have in fighting the dangerous climate conditions – this is obvious if we look at their policies for fighting climate change. California says it will outlaw gasoline cars and trucks by 2035.[62] Meanwhile, California will be spending US$54 billion for fighting climate,[63] including "a measure to prevent the state's last nuclear power plant from closing." California also added "sharp new restrictions on oil and gas drilling." But like in the case of the car reign in California, the new mandate would bring the carbon dioxide era to an end in 2045. However, 2035 and 2045 are five and fifteen years too late. As repeatedly mentioned, Antonio Guterres, UN Secretary General, and UN climate scientists have been warning the world that we must eliminate at least 50% of fossil fuels by 2030. California, harmed by unprecedented heat waves, is not changing their policies fast enough. Farmers still command 80% of the state's diminishing drinking water.[64,65] Millions of large polluting petroleum-powered cars are the kings on California's roads and highways. In 2019, there were about 36.5 million registered vehicles in California.[66]

Oregon, like California, is also in similar situation. About 85% of the drinking water of Oregon goes to farmers.[67] And in 2023, there were about 4.5 million registered vehicles in Oregon.[68] Both states have large forests, so when fires are ignited, for the most part, in the sky, the spectacle and the effects resemble the fire war between the Titans and the gods.

---

[61] David Wallace-Wells, The American west's haunting smoke-filled future, *New York Times*, August 24, 2022.

[62] Soumya Karlamangla, What to know about California's ban on new gasoline-powered cars, *New York Times*, August 29, 2022.

[63] Brad Plumer, California approves a wave of aggressive climate measures, *New York Times*, September 1, 2022.

[64] Jennifer Medina, Farmers agree to water cuts in California, *New York Times*, May 22, 2015.

[65] River's End: California's Latest Water War. PBS, 2021.

[66] State of California, Department of Motor Vehicles, 2019.

[67] US Drought Monitor, Oregon Irrigation Consumptive Use Project.

[68] Oregon Department of Motor Vehicles Registration Statistics, 2023.

Hesiod says the heat of the fire war almost melted the Earth.[69] While the fires in California and Oregon are tiny compared to those extremely ancient conflagrations, they are large, very large in my opinion.

An Oregon ecologist, Timothy Ingalsbee, director of Firefighters United for Safety, Ethics, and Ecology, said that those "very large fires" that burned tens of thousands of acres on the west side of the Cascade Mountains of Oregon, included "screaming winds" from the deserts as they "barrel[ed] down our valleys, marching right up to the doorsteps of major metropolitan areas like Portland and Eugene... the whole region [was] under a pall of smoke. [The smoke was] literally blotting out the sun, from British Columbia to Baja." He spoke with conviction and knowledge that the fires in Oregon were "climate fires." He explained that the fires were a consequence of "extreme heat waves and prolonged droughts and very low humidity." He rightly warned that "no number of firefighters, engines, air tankers, or whatever, will be able to handle phenomena like this... climate-driven wildfire. Nature is far more powerful than us. And so, unless and until we get a handle on our fossil fuel emissions, there's nothing we can do that will really prevent these kinds of events from happening." Ingalsbee also talked about his daughter who has sued the federal government for its support of the fossil fuel industry, which is undermining her life and the lives of millions of young Americans. He said: "It's not the inaction of the federal government that is part of the climate crisis; it's their deliberate actions, pushing more fossil fuel extraction and burning... promoting the alteration of the planet's atmosphere and oceans... the government is causing my daughter's generation and all future generations, of all species, for that matter, significant harm.... All administrations [from both parties] for decades have knowingly been putting the planet in peril by promoting the fossil fuel civilization."[70]

Ingalsbee spoke like a philosopher. His knowledge of ecology and his experience as a fire fighter gave him the wisdom to see the political and fossil fuel connection that brought climate change into being. Both the Republican and Democrat administrations have been subsidizing fossil fuels; and the availability of cheap fossil fuels trigger a warmer planet – threatening states like Oregon and California with the (potential) death of millions of people on the paths of wild and uncontrollable fires, and the world at large with heat waves and droughts intensified by climate

---

[69] Hesiod, *Theogony*, 693–697.
[70] Democracy Now, September 14, 2020.

change. "Do something about fossil fuel emissions," he said. And I concur; with fossil fuels making up 75% of global greenhouse gasses and 90% of all CO_2 emissions,[71] little else will help to get away from the path of this climate catastrophe unless the governments of the United States, China, Russia, the European Union, India and Brazil phase out fossil fuels quickly. President Trump, for example, took the US out of the 2015 Paris climate accord and denies climata change. His administration all but extinguished environmental protection policies. And since his Republican supporters control Congress and the Supreme Court, the government has become a cheerleader for the fossil fuel industry. Moreover, Trump climate denial will disrupt international climate negotiations, with the result global warming will intensify for the planet and humans.

The climatological "events" of Oregon played out in California as well. I believe that the intense heat, dry conditions and large forests coupled with climate change propel wildfires all over the state and the rest of the country, fires that continued in 2021–2025. I cited evidence above that millions of Californians breathed air filled with smog and fire smoke. I was one of those millions. The fires and heat and the unhealthy air left no doubt that California, as Governor Newsom declared, was in the midst of a climate calamity.[72] That calamity continued in 2021 and 2022 – and lingered in slightly less hazardous state in 2023 and 2024. UN Secretary General Antonio Guterres convinced us that the world was not that far behind California and Oregon.

In addition, on August 9, 2021, the hundreds of climatologists of the United Nations published their climate report, warning the international community that climate danger was real, man-made, and affecting all ecosystems in all regions of the planet. Antonio Guterres strongly endorsed the report, saying it's "Code Red for humanity."[73] But even before this dire warning, the 2018 report was so alarming that a UN official described it as "a deafening, piercing smoke alarm going off in the kitchen."[74] Yet the annual UN climate summits continuously demonstrate that similar warnings are not taken seriously by states and business

---

[71] United Nations, Causes and effects of climate change (n.d.). https://www.un.org/en/climatechange/science/causes-effects-climate-change (Accessed: 10 September 2024).

[72] Susanne Rust and Tiny Barboza, California's climate apocalypse, *Los Angeles Times*, September 12, 2020.

[73] "Secretary-General Calls Latest IPCC [Intergovernmental Panel on Climate Change] Report Code Red for Humanity," *UN Press Release*, August 9, 2021.

[74] Editorial, What will success look like in Glasgow? *New York Times*, October 23, 2021.

corporations, particularly fossil fuel companies. It is as if destructive forest fires, drought, famines, environmental refugees, floods, the killing of corals, the extinction of animal species, the burning of about 20% of the centuries-old Sequoias in California,[75] the melting of glaciers, the thawing of permafrost, and rising sea levels and worsening hurricanes are not alarming enough. On hurricanes alone, the news is very bad. Porter Fox, researcher and author on worsening weather, says we are entering "a new era of extreme weather... extreme cyclones... storm intensity." Anthropogenic climate change, he says, has made supercharged hurricanes "a ficture of life around the world, and they are going to get worse – with millions of people in their cross hairs."[76]

Close to 197 countries have repeatedly pledged to limit their greenhouse emissions, but global emissions continue to rise. IPCC report of 2018 described a bleak future for the planet at an increase of the current temperature to 1.5°C above pre-industrial levels.[77] Policy makers the world over have heard these warnings at the climate meetings at Kyoto in 1997, Copenhagen in 2009, and Paris in 2015. However, fighting climate change is fighting fossil fuel companies, which are some of the most powerful companies on Earth. The struggle is unequal, but it is going on. The story below highlights that struggle.

Burning Fossil Fuels Keeps Increasing Global Temperature

A report (dated October 20, 2021) by the UN Environment Program (UNEP) says that "Governments' fossil fuel production plans [are] dangerously out of sync with [their commitments at the] Paris [climate conference of 2015]." In more detail, the UNEP report explains the hidden reality behind the talk of fossil fuel-governments:

> "The world's governments plan to produce around 110% more fossil fuels in 2030 than would be consistent with limiting warming to 1.5°C, and 45% more than consistent with 2°C. The size of the production gap

---

[75] Lila Seidman, Fires take heavy toll on sequoias, *Los Angeles Times*, November 20, 2021.
[76] Porter Fox, Hurricane Milton is terrifying and it is just the start, *New York Times*, October 9, 2024.
[77] Intergovernmental Panel on Climate Change, Special Report, Global Warming of 1.5°C (2018).

has remained largely unchanged compared to our prior assessments. Governments' production plans and projections would lead to about 240% more coal, 57% more oil, and 71% more gas in 2030 than would be consistent with limiting global warming to 1.5°C. Global gas production is projected to increase the most between 2020 and 2040 based on governments' plans. This continued, long-term global expansion in gas production is inconsistent with the Paris Agreement's temperature limits. Countries have directed over USD 300 billion in new funds towards fossil fuel activities since the beginning of the COVID-19 pandemic – more than they have towards clean energy."[78]

Last Chance?

The global climate conference, which lasted from October 31 to November 12, 2021, in Glasgow, Scotland, United Kingdom, was probably a failed effort for the world to put the brakes on climate change affecting both humans and wildlife on this beautiful Earth. Government representatives who were present at the Glasgow conference appeared to disregard human and environmental health; They decided to continue with the current burning of fossil fuels. They – the organizers of the Glasgow climate summit reflecting the politics of prime ministers and presidents – ignored the emissions from agriculture, which are substantial. We cited evidence above that they range from 24% to around 51% of the total greenhouse emissions entering the atmosphere from all over the world. On October 26, 2021, the US Environmental Protection Agency reported 24% of the total US greenhouse emissions came from agriculture. A November 8, 2021 study[79] of the data collected from 236 countries and territories by the UN Food and Agriculture Organization showed that, in 2019, industrialized agriculture and food processing, packaging, and transport were responsible for 16.5 billion metric tons of greenhouse gas emissions – 31% of the total world anthropogenic emissions causing climate chaos. Of the 16.5 billion tons, some 7.2 billion tons originated

---

[78] UN Environment Program, Governments' fossil fuel production plans dangerously out of sync with Paris limits, October 20, 2021.

[79] Francesco N. Tubiello *et al.,* Pre- and post-production processes along supply chains increasingly dominate GHG [greenhouse gas] emissions from agri-food systems globally and in most countries, *Earth System Science Data*, November 8, 2021; Natasha Gilbert, One-third of our greenhouse gas emissions come from agriculture, *Nature*, October 31, 2012.

within farms, 3.5 billion tons came from land use, and 5.8 billion tons from food processing, packaging, and transport. In 2019, deforestation was the "largest source" of greenhouse gas emissions, closely followed by livestock manure, the eating of food, food waste, and the petroleum and natural gas burned by farmers and food retailers. If anything, I think that these numbers probably underestimate the global warming legacy of agriculture.

The Glasgow climate conference was destined to fail, and it did. It failed to live up to the expectations of billions of people – and thousands of scientists. Two hundred of those scientists signed a letter they sent to the presidents and prime ministers in Glasgow. They pleaded and urged them to take "immediate, strong, rapid, sustained and large-scale actions" to prevent climate catastrophe.[80] The climate meeting in Glasgow, the scientists suggested, decided the fate of billions, including the fate of the planet. In a last-ditch effort, the UNEP warned the presidents and prime ministers at Glasgow that their pledges did little to confirm the promises they had made in 2015 in Paris to keep the temperature of the planet at 1.5°C above pre-industrial levels. The UNEP said that the new pledges nations made at Glasgow in early November 2021 would "put the world on track" for a 2.7°C increase by the end of the century, which guarantee "catastrophic changes to the Earth's climate."[81]

Secretary General of the UN, Antonio Guterres, was afraid of that. He knew the latest findings of climatologists were alarming, but he also knew that it would take a lot more to convince world leaders of its importance. A month before Glasgow, he warned: "We can either save our world or condemn humanity to a hellish future."[82] However, Guterres' worries came true. The world leaders appeared not convinced. While they pledged to achieve net zero emissions and reduce methane emissions during the conference, countries such as India, China, and the US continued burning coal with the phase down coal commitment instead of phrasing out coal. Russia and most of the world continued extracting and burning petroleum. To be fair, the world leaders at the conference likely knew the dangers in the continual use of fossil fuels. As I will explain below, clean energy is

---

[80] Scientists appeal for immediate climate action at COP 26, *France 24*, November 11, 2021.

[81] UNEP and UNEP Copenhagen Climate Center, Emissions Gap Report 2021.

[82] UN Chief Antonio Guterres warns of Hellish Future Ahead of Key Climate Summit, *NDTV*, October 1, 2021.

trying to catch up and surpass the energy from fossil fuels. In 2021, Elizabeth Kolbert, reporter of the *New Yorker Magazine*, said, "when it comes to climate change, there's no making up for lost time. Every month that carbon emissions remain at current levels – they're running at about forty billion tons a year – adds to the eventual misery."[83]

The Planet Is Screaming at Us

Americans should understand Antonio Guterres' warnings. They can see it in the burning of the West Coast by gigantic forest fires, the extensive drought in California, as well as the flooding and destruction of the Gulf Coast by cyclones in 2020,[84] the deep freeze of Texas in 2021,[85] and the flooding of the Northeast – especially New York City – in early September 2021.[86] Hurricane Sally brought catastrophic flooding to the southern states. Roads "turned to rivers." The hurricane "dumped a torrent of rain." The National Hurricane Center spoke about "catastrophic and life-threatening flooding." In February 2021, blackouts occurred, taps ran dry, gas pipes froze, water treatment plants failed, and grocery stores ran out of food – in huge Texas.[87] Hurricane Ian made a landfall in Southwestern Florida in September 2022 (Fig. 4). Its lacerating winds with speeds of up to 155 miles per hour shreded infrastructure, pulverized roads, toppled trees, and made homes and businesses "heaps of wood pulp and broken concrete." Losses reached as much as US$40 billion.[88]

Climate change magnified enormously these disasters.[89,90] In addition, in October 2021, the US national security agencies said climate change

---

[83] Elizabeth Kolbert, Against the clock, *The New Yorker*, November 15, 2021.

[84] Hundreds rescued as floods from hurricane sally hit Florida and Alabama, *New York Times*, September 16, 2020.

[85] Nicholas Bogel-Burroughs, Giulia McDonnell Nieto del Rio and Azi Paybarah, Texas Winter Storm: What to Know, *New York Times*, February 20, 2021.

[86] Rachel Ramirez and Drew Kann, New York City was never built to withstand a deluge like the one Ida delivered, *CNN*, September 5, 2021.

[87] Avie Schneider, Several days into the Texas deep freeze, food is scarce, *NPR*, February 18, 2021.

[88] Patricia Mazzei *et al.*, Hurricane Ian's staggering scale of wreckage becomes clearer in Florida, *New York Times*, September 29, 2022.

[89] Hundreds rescued as floods from Hurricane sally hit Florida and Alabama, *New York Times*, September 16, 2020.

[90] National Weather Service: Hurricane Sally – NWA 2021 Research Information.

Fig. 4.   Hurricane Ian over Florida, September 28, 2022.
Source: NASA.

was sapping the strength of the nation.[91] They warned of global food shortages, unrests over land and water, ice melting in the Arctic, and fierce competition over fish and the mining of minerals. They predicted there would be about 140 million environmental refugees in South Asia, Africa, and Latin America. Federal and state regulators warned of climate chaos threats to the economy: floods, hurricanes, and forest fires destroying property worth billions.

The destruction from hurricanes keep spreading and the costs are rising because local governments and the federal government are unwilling to prevent the overpopulation of Florida and other attractive tourists destinations in southeastern states. Building homes and apartments in the wetlands and mangroves of these regions is unsafe as ecological protection diminishes and the harm from storms, intense rains, and hurricanes increases.[92]

It would seem that the Republicans in Congress are not convinced about climate change. The result of a divided Congress and country disrupts its minimal efforts to fight the existential threat of a warming planet.

---

[91] Christopher Flavelle, Julian E. Barnes, Eileen Sullivan, and Jennifer Steinihouser, Climate change poses a widening threat to national security, *New York Times*, October 21, 2021.

[92] Ian Prasad Philbrick and Ashley Wu, Population growth is making hurricanes more expensive, *New York Times*, December 2, 2022.

From the US$1 trillion infrastructure bill that passed Congress in early November 2021, with most Republicans opposing it, only US$47 billion was allocated for climate "resilience." In other words, this money will be spent in the next 10 years to prepare the country to withstand extreme fires, floods, hurricanes, and droughts.[93]

The good news, however, was that the House of Representatives and the Senate voted legislation approving US$369 billion for fighting climate chaos by subsidizing electric cars, solar panels, wind farms, and the country's adoption of green energy.[94] I have hopes that the country is not entirely asleep. "The science," said (then) President Biden, "is clear, we only have a brief window left before us to raise our ambitions and to raise – to meet the task that's rapidly narrowing… But… within the growing catastrophe, I believe there's an incredible opportunity. Not just for the United States, but for all of us. We're standing at an inflection point in world history, we have the ability to invest in ourselves and build an equitable clean energy future and in the process create millions of good-paying jobs and opportunities around the world – cleaner air for our children, where bountiful oceans, healthier forests and ecosystems for our planet."[95]

Despite those brave words, President Biden and his Secretary of the Interior, Deb Haaland, appeared to have forgotten their promises to fight climate chaos shortly afterwards. Not many days after the Glasgow climate summit, where President Biden displayed an admirable passion for truth, he authorized 80 million acres in the Gulf of Mexico/Gulf of America for oil drilling. It is possible that this expansion of offshore drilling was the doing of the Trump (2017–2021) administration and was ordered by a judge. But why not appeal such a decision or issue a moratorium on offshore drilling? Simultaneously, and in conflict with his promises to do away with drilling in public lands, he continued the usage of public lands for private petroleum profit. The Department of Interior issued a report, in which it proposed increasing the fees for leasing public lands and offshore water for drilling. Climate change was unfortunately not mentioned in this government report. The concern of President Biden

---

[93] Coral Davenport and Christopher Flavelle, Infrastructure Bill Makes First Major U.S. Investment in Climate Resilience, *New York Times*, November 6, 2021.

[94] H. R. 5376 – Inflation Reduction Act of 2022 or Public Law No: 117–169.

[95] Coral Davenport, House Passes the Largest Expenditure on Climate in U.S. History, *New York Times*, November 19, 2021.

and Secretary Haaland was simply to "provide a fair return to taxpayers."[96] No wonder the Secretary of the Interior, whose staff authored this report, was accused of ignoring the US Geological Survey's warning that 25% of all US greenhouse gas emissions come from drilling on public lands and federal waters for petroleum.[97]

Camilo Mora, professor of geography and the environment at the University of Hawaii, explained: "The planet is screaming to us... When are we going to start listening?"[98] Many other scientists share the anger and rightness of Mora. On October 7, 2021, they wrote a letter[99] to President Joe Biden, imploring him to bring the fossil fuel era to an end.[100]

> *As U.S. scientists, we write with the utmost alarm about the state of our climate system. As overwhelming evidence shows, the reality of our situation is now so dire that only a rapid phase-out of fossil fuel extraction and combustion can fend off the worst consequences of the climate crisis. In this time of peril, we call on you to fulfill your campaign pledge to listen to science, take bold action to lead the nation back from the brink of runaway climate chaos, and lead the world in a rapid transition away from fossil fuels. We urge you to:* **Completely stop federally authorized fossil fuel expansion** *by banning new federal fossil fuel leasing, fracking, and drilling on public lands and waters...* **Declare a climate emergency** *to reinstate the decades-long ban on crude oil exports, direct a portion of military spending to a rapid construction program of renewable energy projects, and provide loan guarantees to advance a rapid [and] just buildout of clean renewable energy;* **Abandon industry delay tactics...** *The people of the United States are already experiencing intensifying climate disasters caused primarily by [the] extracting, transporting, and burning [of] fossil fuels, and the types of disasters we*

---

[96] US Department of the Interior, Report on the Federal Oil and Gas Leasing. November 2021.

[97] Coral Davenport and Lisa Friedman, Interior Dept. report on drilling is mostly silent on climate change, *New York Times*, November 26, 2021.

[98] John Schwartz, A season of climate-fueled disasters, *New York Times*, September 16, 2020.

[99] An open letter from U.S. scientists imploring president Biden to end fossil fuel era, October 7, 2021.

[100] Center for biological diversity: An open letter from US scientists imploring president Biden to end fossil fuel era, October 7, 2021.

are living through now are certain to get worse – much worse – without emergency action.... UN Secretary-General António Guterres has declared the newest findings of the recent IPCC report a code red for humanity. We urge you to convene world leaders around a shared vision for moving humanity off fossil fuels and avoiding carbon offset schemes and fossil-fuel industry delay tactics. Our chances for avoiding irreversible and uncontrollable climate chaos diminish daily. We implore you, on behalf of and for the love of all life on Earth, to respond to the greatest threat ever to face our species and lead the transition away from fossil fuels that humanity desperately needs. (Emphasis mine)

President Biden and Secretary Haaland probably did not respond to this dramatic appeal from American scientists. These scientists really understand the implications of humans awakening the anthropogenic climate crisis. The Republicans and their billionaire funders, some Democratic politicians, and most corporations are not about to listen to reason, either. The UN Chief, Antonio Guterres, agreed with Camilo Mora and other scientists, likely because he has had intimate knowledge of the agenda of the fossil fuel lobbyists and industry. About three weeks after the scientists sent their letter to Biden, Guterres addressed those heads of state attending the world climate summit in Glasgow, Scotland. He said to them:

The six years since the Paris Climate Agreement [of 2015] have been the six hottest years on record. Our addiction to fossil fuels is pushing humanity to the brink. We face a stark choice: Either we stop it – or it stops us. It's time to say: enough. Enough of brutalizing biodiversity. Enough of killing ourselves with carbon. Enough of treating nature like a toilet. Enough of burning and drilling and mining our way deeper. We are digging our own graves.[101]

Guterres is a courageous and moral man. I think that he honors his high office and serves the interests of humanity and those of the Earth. He knows the truth about climate because he has the world's best climatologists working for him. In comparison, I believe that the same cannot be said of the presidents, prime ministers, chief business executives, and

---

[101] Secretary-General's remarks to the World Leaders Summit – COP 26, *UN News*, November 1, 2021.

billionaires or oligarchs who influence the policy-making of the countries of the world. Thucydides, the great Athenian historian of the fifth century BCE, tells us that the political leader of Athens, Pericles, was unhappy with any Athenian pretending he did not like politics because he minded his own business. That, Pericles said, was unacceptable for a citizen. Such a person, Pericles said, did not fit in the polis, and lived a useless life.[102] A generation after Pericles and Thucydides, in the fourth century BCE, Aristotle, great philosopher-scientist and tutor of Alexander the Great, said that man is by nature a political animal, the only one nature endowed with the virtue of speech and wisdom, which enable him to create a political community, a polis that makes possible security and civilization. This understanding of the bad and the good, just and unjust, is part of nature, which is perfect, Aristotle said. But take the virtues of law and justice away from man, according to Aristotle, and man becomes a vicious monster: that is, a man who is utterly devoid of goodness. Such a man is the most unscrupulous and savage of animals.[103]

I am sure that presidents and prime ministers have heard of Aristotle and may even understand the crucial role of virtue in human behaviors and politics. But at the current state of the world, it appears to me, that they are primarily people of business, perhaps even forgetting that children and young people are very much threatened by a warming planet. During the COP26 held in Glasgow in 2021, youth activists put up a vigorous protest outside the conference. They denounced the irresponsible, cowardly and dangerous inaction of world leaders during their protest, shouting:

> *"No more blah blah blah!"*
> *"Climate justice now!"*
> *"Stop extraction and protect indigenous communities!"*
> *"Protect environmental defenders!"*
> *"You'll die of old age; we'll die of climate change!"*
> *"Code red for humanity!"*[104]

Thankfully, some Americans are listening. Jane Fonda is one. She is a famous actress and political dissident. She is urging people to wake up

---

[102] Thucydides, *The Peloponnesian War* 2.40.

[103] Aristotle, *Politics* 1253a.

[104] The youth aren't having it with leaders' inaction, *UN News*, November 5, 2021.

to the reality of climate change. "It's a dire situation that we're facing," she says, "and we have little time left to really do what's needed, what the science tell[s] us we have to do. The reason we have so little time is because the fossil fuel industry lied to us 40-plus years ago. They knew what they were doing. Their scientists told them that they were causing global warming and that it was going to be catastrophic for the world. And they did it anyway and lied to us and tried to make us doubt the science. And, as a result, the window has shrunk in which we can do something."[105] Fonda's wisdom mirrors the latest climatological evidence. She is telling Americans they have to resist the politicians and politics of oil billionaires. Like pesticides, fossil fuels have to be removed. Yes, no more petroleum, natural gas and coal. But we also need to change our mentality and way of life very quickly. Fonda's straight talk helps. She urges Americans to rethink the hazardous present and suggested that the United States could provide a more inclusive future that embraces people of different skin colors. In addition – and this is absolutely necessary – Fonda believed that Americans need to understand why climate change is the result of burning fossil fuels.

Fonda's warnings mirror Antonio Guterres' warnings. And Antonio Guterres' warnings come straight from the horses' mouths: the reports of the IPCC. In late March 2022, Guterres warned that, "The goal to limit future warming to 1.5°C, highlighted in the 2015 Paris Agreement on climate change, and driven home in last November's COP26['s] gathering in Glasgow, is now on 'life support' and 'in intensive care.'... Instead of 'hitting the brakes' on the decarbonizing the global economy, 'put the pedal to the metal towards a renewable energy future.'"[106]

Written by the world group of scientists working under the auspices of Antonio Guterres and the United Nations, the 2022 report[107] paints a bleak picture of the current state in fighting climate change. The Chair of the Panel said that the April 4, 2022 report, as well as all other IPCC reports, should be taken seriously. Climate change is threatening the planet, people, and civilization: "The preceding IPCC reports are clear – human-induced climate change is widespread, rapid, and intensifying. It is a threat to our well-being and all other species. It is a threat to the health

---

[105] Jane Fonda on taking action to address 'dire' climate crisis, *PBS Newshour*, September 10, 2020.

[106] Antonio Guterres, Sleepwalking to Climate Catastrophe, *UN News*, March 21, 2022.

[107] Intergovernmental Panel on Climate Change, *Climate Change*, 2022.

of our entire planet. Any further delay in concerted global climate action will miss a rapidly closing window."[108]

Antonio Guterres, however, lashed out against the deception and lies of corporations and governments. He accused them of hypocrisy and setting the planet on fire. "The jury has reached a verdict. And it is damning," he said. "This report of the Intergovernmental Panel on Climate Change is a litany of broken climate promises. It is a file of shame, cataloguing the empty pledges that put us firmly on track towards an unlivable world. We are on a fast track to climate disaster. Major cities under water. Unprecedented heatwaves. Terrifying storms. Widespread water shortages. The extinction of a million species of plants and animals. This is not fiction or exaggeration. It is what science tells us will result from our current energy policies. We are on a pathway to global warming of more than double the 1.5°C [Celsius] limit agreed [on] in Paris [in 2015]. Some Government and business leaders are saying one thing but doing another. Simply put, they are lying. And the results will be catastrophic. This is a climate emergency. Climate scientists warn that we are already perilously close to tipping points that could lead to cascading and irreversible climate impacts. But, high-emitting Governments and corporations are not just turning a blind eye, they are adding fuel to the flames. They are choking our planet, based on their vested interests and historic investments in fossil fuels, when cheaper, renewable solutions provide green jobs, energy security and greater price stability… the truly dangerous radicals are the countries that are increasing the production of fossil fuels."[109]

Unfortunately, the state of climate in 2023 was even worse than that of 2022. The IPCC's 2023 report, like its predecessors, told it the way it is. The jargon was minimal and the dire message overwhelming:

> *Human activities, principally through emissions of greenhouse gases, have unequivocally caused global warming, with global surface temperature reaching 1.1°C above 1850–1900 in 2011–2020. Global greenhouse gas emissions have continued to increase, with unequal*

---

[108] United Nations, *Remarks by the IPCC Chair during the Press Conference presenting the Working Group III contribution to the Sixth Assessment Report,* April 4, 2022.

[109] United Nations, *Secretary-General Warns of Climate Emergency, Calling Intergovernmental Panel Report "a File of Shame," While Saying Leaders "Are Lying" Fuelling Flames,* April 4, 2022.

historical and ongoing contributions arising from unsustainable energy use, land use and land-use change, lifestyles and patterns of consumption and production across regions, between and within countries, and among individuals... Widespread and rapid changes in the atmosphere, ocean, cryosphere [frozen ice regions], and biosphere have occurred. Human-caused climate change is already affecting many weather and climate extremes in every region across the globe... Risks and projected adverse impacts and related losses and damages from climate change escalate with every increment of global warming. Climatic and non-climatic risks will increasingly interact, creating compound and cascading risks that are more complex and difficult to manage... Climate change is a threat to human well-being and planetary health. There is a rapidly closing window of opportunity to secure a livable and sustainable future for all. Climate resilient development integrates adaptation and mitigation... The choices and actions implemented in this decade will have impacts now and for thousands of years.[110]

The UNEP also published the 2023 IPCC report[111] and wholeheartedly supported its conclusions. UNEP correctly highlighted the impact of climate change on cities where most of the world's people live. "Climate change," UNEP said, "is a global phenomenon that largely impacts urban life. Rising global temperatures causes sea levels to rise, increases the number of extreme weather events such as floods, droughts, and storms, and increases the spread of tropical diseases. All these have costly impacts on cities' basic services, infrastructure, housing, human livelihoods, and health. At the same time, cities are a key contributor to climate change, as urban activities are major sources of greenhouse gas emissions. Estimates suggest that cities are responsible for 70% of global CO_2 emissions, with transport and buildings being among the largest contributors."[112]

---

[110] Intergovernmental Panel on Climate Change, *Climate Change 2023: AR6 Synthesis Report.* https://www.ipcc.ch/report/ar6/syr/.

[111] United Nations Environment Program, *Climate Change 2023: Synthesis Report.* https://www.unep.org/resources/report/climate-change-2023-synthesis-report.

[112] United Nations Environment Program, *Cities and Climate Change.* https://www.unep.org/explore-topics/resource-efficiency/what-we-do/cities-and-climate-change.

Setting Civilization and Mother Earth on Fire

No doubt, both the 2023 IPCC and UNEP reports are right. Continuing to burn fossil fuels adds to the frequencies of forest fires, drought, catastrophic floods, and asphyxiating dead zones in seas and oceans. In effect, continuing the consumption of petroleum, natural gas, and coal "as usual" is akin to setting both civilization and Mother Earth on fire. These human and natural phenomena make modern civilization extremely vulnerable. In fact, big oil is threatening civilization and life on Earth.[113]

I feel that it is worth repeating Guterres' warning that governments are lying to their people about climate change threatening them. World leaders are not doing enough to reverse the effects of climate change. This behavior is not getting all of us anywhere. The Earth is billions of years old while humans and civilization have only existed on Earth for a few hundred thousand years. Humans may find out that extinction is the price of not listening to the facts of science. Agriculture hides some of those facts.

Hidden and Real Costs of Factory Farming

The UN Food and Agriculture Organization (FAO) finally started sharing the facts regarding the destructive effects of industrialized agriculture in 2023. "Although current agrifood systems," FAO said, "provide nourishment and sustain economies, they also impose huge hidden costs on health and the environment – the equivalent of at least US\$10 trillion annually." The FAO report highlighted that the largest "hidden costs," close to as much or more than 70%, are a result of bad diets "high in ultra-processed foods, fats and sugars, leading to obesity and noncommunicable diseases." Moreover, FAO underlined that such unhealthy practice is prevalent in "richer countries."[114]

FAO underestimated the costs, in my opinion. It calculated that "one fifth of the total costs are environment-related, from greenhouse gas and nitrogen emissions, land-use change and water use, with all countries affected." Without details, we do not know how FAO reached these estimates. In my viewpoint, with factory agriculture alone, the ecological and human health costs are immense.

---

[113] Evaggelos Vallianatos, Big oil and civilization don't mix, *Counterpunch*, May 17, 2024.
[114] UN Food and Agriculture Organization, *State of Food and Agriculture 2023*. https://doi.org/10.4060/cc7724en.

Industrialized farming is applied petroleum. The tractors and other gigantic machinery rely on petroleum. Pesticides are called petrochemicals because they are varieties of petroleum. Synthetic fertilizers are mixtures of chemicals like nitrogen, phosphorus, potassium, natural gas, and ammonia. These synthetic chemicals are usually contaminated by toxic metals like lead, mercury, arsenic, and cadmium. Farmers also use fertilizers coming from city sludge. These fertilizers are contaminated by forever chemicals (or PFAS – per- and polyfluoroalkyl substances) harming and poisoning the soil, the food, water and air.[115] Altogether, these synthetic and sewage fertilizers spread on millions of acres of fertile land for decades, harm the food (crops, fruits, and vegetables), and the health of the people who eat such industrialized food. Besides synthetic fertilizers, insecticides and weed killers also harm the soil, wildlife, and the environment. These facts about fertilizers and chemical sprays tell me that the FAO report grossly under-reported the hidden and real costs of agribusiness and agrifood.

Any Solutions?

The 2023 IPCC report was another warning to all of us, but especially to world leaders, that humanity is at risk. Presidents, prime ministers and owners of petroleum corporations showed up at Dubai during December 2023, explaining the "progress" they made against climate change. My guess is that the UN Chief, Antonio Guterres, denounced them once again. Yet calling out the power holders with epithets has its limits. In my opinion, petroleum-powered wars in Ukraine, Israel and Palestine and Lebanon are signs that war could potentially spread to other countries or to the entire world. Unfortunately, it appeared to me that the United States and its NATO allies are funding these wars. Since the early 1990s, there were suggestions that the neoconservatives in America preferred global hegemony over the supremacy of the UN Charter.[116] I believe that this intensifies the crises of a warmer planet in today's world.

The reliable climate science in the pages of the 2023 IPCC report leaves no doubt that human activities are responsible for the cosmic

---

[115] Hiroko Tabuchi, Something's poisoning America's Land. Farmers fear forever chemicals, *New York Times*, August 31, 2024.

[116] Jeffrey D. Sachs, How the neocons chose hegemony over peace beginning in the early 1990s, *Racket News*, September 4, 2024.

tragedy of dangerously heating up the planet. This planet, Mother Earth, the nursery of life in the universe, belongs to all of us. Time has come for us to pay attention to what each of us and our politicians do about climate change. I would suggest the following: Put solar panels on the roof of your house and demand your city adds solar panels on its buildings and over parking lots. Demand that your city, state and the federal government support and fund public transportation, electric cars, electric buses, and electric trams – all over the cities and all over the country. Stop electing people who are apathetic or neglectful or deniers of the climate emergency. These are life and death decisions I feel.

The United States Getting Dangerously Warm

Guterres and the 2023 IPCC report painted a bleak picture. The world is in an emergency situation from the effects of hurricanes, droughts, heat waves, on land and seas, melting of ice, and thowing of the permafrost. These natural but anthropogenic climate effects are more severe on America because of the country's almost complete inundation by fossil fuels for more than a century. America is like another planet. In my opinion, perpetual wars, overconsumption, super individualism, and the folly of endless economic growth go hand in hand with the ceaseless burning of fossil fuels. The price America is paying is high. Since 2011, about 1.5 trillion gallons of water under intense pressure facilitated fracking for extracting petroleum and natural gas in America. Reporters for the *New York Times* reveal the harm of using drinking water for fracking. This "insatiable search for oil and gas," they write, "has become the latest threat to the country's endangered aquifers, a critical national resource that is already being drained at alarming rates by industrial farming and cities in search of drinking water."[117] In a country where drinking water remains a concern, I feel that this massive waste of water for fracking remains an invisible subsidy to companies ripping apart the land. More on subsidies below.

Ironically, in 2022, the US government circulated a draft climate report, for comment.[118] According to the *New York Times*, the draft report

---

[117] Hiroko Tabuchi and Blacki Migliozzi, Monster Fracks Are Getting Far Bigger and Far Thirstier, *New York Times,* September 25, 2023.

[118] US Global Change Research Program, National Climate Assessment, Draft, 2022.

said "the effects of climate change are already far-reaching and worsening throughout all regions in the United States, posing profound risks to all aspects of society; from drinking water supplies in the Midwest, to the small businesses [operating] in the Southeast." But the most dire conclusions of the draft report suggested that "the things Americans value most are at risk… More intense extreme events and long-term climate changes make it harder to maintain safe homes and healthy families, reliable public services, a sustainable economy, thriving ecosystems and strong communities."[119] In a report by CNBC, the National Climate Assessment draft also highlighted that "the effects of climate change are felt most strongly by communities that are already overburdened, including indigenous people, people of color, and low-income communities." Unfortunately, these affected communities are often the least responsible for the greenhouse gas emission that cause climate change."[120]

The results are extremely unpleasant. No American can deny the impact of climate change on their daily life. Those living in Miami Beach, Florida, for example, can hardly deny the increased frequency of flooding, which we discussed above. In Alaska, climate change brought marine heat waves that destroyed 14 fisheries. In Colorado, insufficient snow nearly wrecked the state's ski industries. I read the draft US government climate change Report as well as the final Report,[121] released on November 14, 2023. They confirm the above-mentioned disasters and sum up the additional climate blows the country will suffer from its continuous use of fossil fuel and not making energy and institutional changes fast enough. "The effects of human-caused climate change," the final Report said, "are already far-reaching and worsening across every region of the United States… without deeper cuts in global net greenhouse gas emissions and accelerated adaptation efforts, severe climate risks to the United States will continue to grow." No doubt, the risks people already face keep worsening. The final Report Assessment warned Americans to expect "increases in heat-related illnesses and death, costlier storm damages, longer droughts that reduce agricultural productivity and stream water

---

[119]Brad Plumer and Raymond Zhong, Draft Report Offers Starkest View of U.S. Climate Threats, *New York Times*, November 8, 2022.

[120]Emma Newburger, Climate change threatens to destroy the things Americans value most, U.S. government warns, *CNBC*, November 8, 2022.

[121]US Global Change Research Program, *Fifth National Climate Assessment*, November 14, 2023. https://nca2023.globalchange.gov.

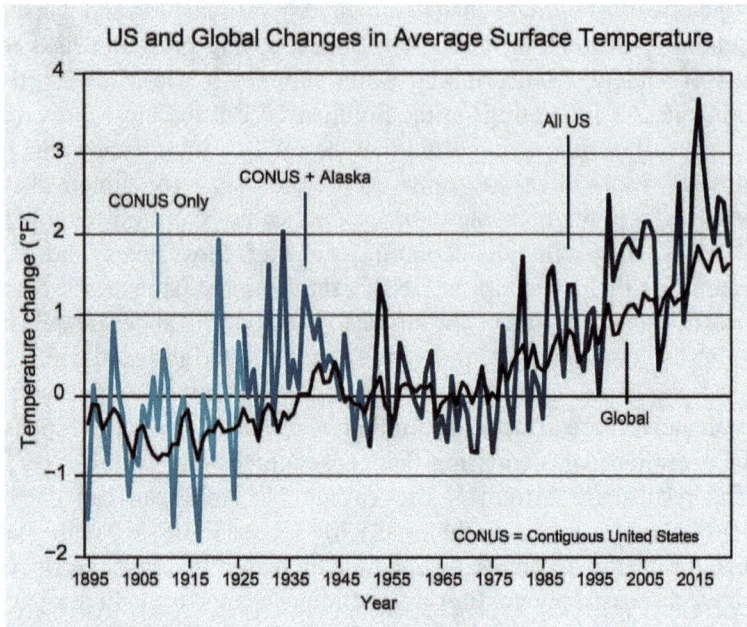

Fig. 5. US and global temperature, 1895–2015. Light blue, US temperature, and dark line, global temperature.

Source: US Fifth Climate Assessment, November 14, 2023; Sachi Kitajima Mulkey, Claire Brown and Mira Rojanasakui, The World Is Warming Up. And It's Happening Faster, *New York Times*, June 26, 2025.

systems, and larger, more severe wildfires that threaten homes and degrade air quality." In addition, the Assessment reported that the United States "experiences, on average, a billion-dollar weather or climate disaster every three weeks." Startling as this revelation is, I never expected to hear that, "Over the past 50 years, the U.S. has warmed about 68% faster than the planet as a whole," – Just imagine the implications of this critical reality.

The *Fifth National Climate Assessment* said that the fast warming of America started in full force in 1972. That was the same year I received my Ph.D. from the University of Wisconsin. Not that my education made any difference in the dramatic changes taking place on Earth and its atmosphere, the latter of which was being blasted daily by unfathomable amounts of greenhouse gases. The significance of 1972 was that the country was then fighting its endless wars in southeast Asia, which it continued in the Middle East and Afghanistan, and, starting in early 2022, in

Ukraine, and in October 2023, in Israel. As we already mentioned, America has been alleged as the funder of the last two wars. We should have known that war is an explosive charge to climate chaos. It accelerates the warming of the planet, boosting the production and burning of fossil fuels.[122] For example, Israel's war on Gaza and the Middle East, funded by the United States,[123] has quite an impact on climate.[124] The same is true of the war in Ukraine and other wars. Yet international conventions like the 1997 Kyoto Protocol and the 2015 Paris Climate Summit excused militaries from reporting their greenhouse gas emissions. The untested assumption is that the militaries of the planet emit about 5.5% of the global carbon emissions.[125]

It also appears that munitions manufacturers are taking advantage of the wars in Ukraine and Israel. The approved increase of war spending in the military budget brings immense profits.[126] In early December 2022, the US House of Representatives passed a gigantic military budget of US$858 billion which, among other things, would keep American weapons flowing to Ukraine and Taiwan.[127]

Add to the military-generated warming of America, the vast industrialized agriculture of the country, the equally vast numbers of petroleum-powered cars, trucks, leaf-blowers, airplanes, ships, yachts, and the fossil fuel-powered production of electricity, cement, fertilizers, and pesticides, and you get an idea of the enormous fossil fuel energy used daily in America. You would understand why America is warming so quickly and so much more than the rest of the countries around the world. To add insult to injury, cities light up their skyscrapers, hotels, and theaters

---

[122] Vallianatos, Climate Change and Wars Are Breaking Down the Foundations of Civilization, *Counterpunch*, January 15, 2024. See also: Neta C. Crawford, "Costs of War," June 12, 2019, Watson Institute, Brown University.

[123] Watson Institute for International and Public Affairs, Brown University, United States Spending on Israel's Military Operations and Related US Operations in the [Middle East] Region, October 7, 2023 – September 30, 2024.

[124] Emissions from Israel's war in Gaza have immense effect on climate catastrophe, *The Guardian*, January 9, 2024.

[125] Sarah Mcfarland and Valerie Volcovici, Insight: World's war on greenhouse gas emissions was a military blind spot, *Reuters*, July 10, 2023.

[126] Jasper Jolly and agencies, Arms maker BAE Systems makes record profit amid Ukraine and Israel-Gaza wars, *The Guardian*, February 21, 2024.

[127] Catie Edmondson, House Passes $858 Billion Defense Bill, *New York Times*, December 8, 2022.

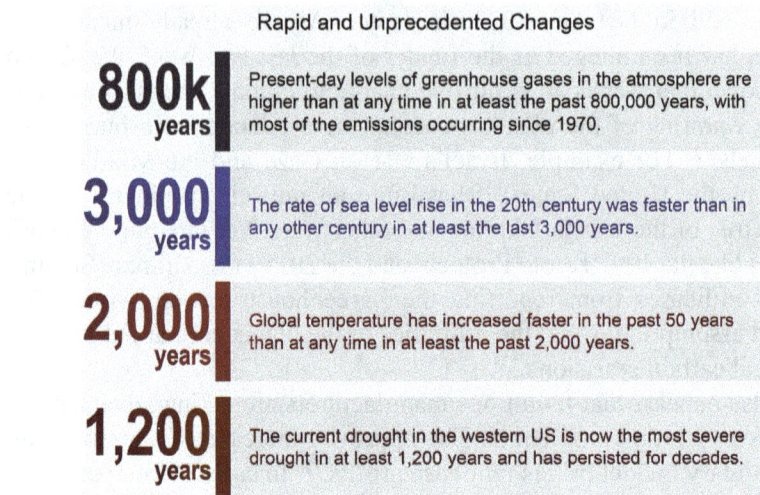

Rapid and Unprecedented Changes

800k years — Present-day levels of greenhouse gases in the atmosphere are higher than at any time in at least the past 800,000 years, with most of the emissions occurring since 1970.

3,000 years — The rate of sea level rise in the 20th century was faster than in any other century in at least the last 3,000 years.

2,000 years — Global temperature has increased faster in the past 50 years than at any time in at least the past 2,000 years.

1,200 years — The current drought in the western US is now the most severe drought in at least 1,200 years and has persisted for decades.

Fig. 6. Rapid and dangerous consequences of the anthropogenic climate chaos – the last 50 years.

Source: US Fifth National Climate Assessment, November. 14, 2023; Henry Fountain, How bad is the western drought? Worst in 12 centuries, *New York Times*, February 14, 2022.

at night. This light pollution is an additional but thoroughly unnecessary blow against the planet.

The *US Fifth National Climate Assessment* and numerous UN and IPCC reports suggest that we know enough about climate change. Yes, we know that the land warms faster than the ocean, that higher latitudes have been warming faster than lower latitudes and that "the Arctic has warmed fastest of all." In addition, some eight disasters costing on the average of about a billion dollars damage each have struck the US every year in the past forty years. However, climate change keeps multiplying these calamities.[128]

Warning from the World Meteorological Organization

The World Meteorological Organization (WMO) reached similar conclusions with the bad news of the *Fifth National Climate Assessment*. It issued its climate warning on November 14, 2023. "Greenhouse Gas

---

[128]David Gellis, Climate disasters daily? Welcome to the new normal, *New York Times*, July 10, 2023.

levels in the atmosphere," said WMO, "have reached record levels. Again. This will continue to trap heat and drive climate change for many years, with more extreme weather, sea level rise and many other impacts on our planet." Petteri Taalas, Secretary-General of WMO, said that in 2022, the most important heat-trapping gas in the atmosphere, carbon dioxide, CO_2, was 50% higher than its pre-industrial levels. Methane and nitrous oxide, the other two important greenhouse gases in the atmosphere, also experienced "the highest year-on-year increase on record from 2021 to 2022." Then the exasperated Taalas said, "Despite decades of warnings from the scientific community, thousands of pages of reports and dozens of climate conferences, we are still heading in the wrong direction." This means temperatures are still exceeding the targets the international community agreed to in 2015 in the Paris Agreement in order to avoid global warming of higher than 1.5°C. The consequences of this failure, Taalas said, would be "more extreme weather, including intense heat and rainfall, ice melt, sea-level rise and ocean heat as well as acidification. The socioeconomic and environmental costs will soar." Taalas explained that about half of the CO_2 emissions stay in the atmosphere, about 25% drop into the seas and oceans, and the rest are absorbed by forests and other ecosystems. We need to remember that the more CO_2 we dump into the atmosphere, the higher the temperatures of the planet. CO_2 has also a "long life" in the atmosphere, so its mischief of global warming will last for decades – even if we reduce CO_2 emissions to zero. "The last time the Earth experienced a comparable concentration of CO_2," Taalas said, "was 3 to 5 million years ago, when the temperature was 2 to 3°C warmer and sea level was 10 to 20 meters higher. There is no magic wand to remove the excess carbon dioxide from the atmosphere."[129]

Climate Change Is Spreading All over the Planet

Petteri Taalas of WMO made clear that conditions in the rest of the world are not that much better than the fierce effects of ceaseless global warming in the over-heated United States. I cited Taalas above saying "there is no end in sight to the burning of fossil fuels." It is no wonder that on November 7, 2022, the UN champion for climate sanity, Antonio Guterres,

---

[129] World Meteorological Organization, No end in sight to rising greenhouse gas emissions, *UN News*, November 15, 2023. https://news.un.org/en/story/2023/11/1143607.

said to the 110 prime ministers and presidents who showed up for the opening of the 27th UN climate meeting in the Egyptian Red Sea resort town of Sharm el-Sheikh that their indifference to the rising climate chaos bordered on the suicidal. He said to the Excellencies:

> *We are in the [climate] fight of our lives. And we are losing. Greenhouse gas emissions keep growing. Global temperatures keep rising. And our planet is fast approaching tipping points that will make climate chaos irreversible.* **We are on a highway to climate hell** *with our foot still on the accelerator.... climate change is... the defining issue of our age. It is the central challenge of our century. It is unacceptable, outrageous, and self-defeating to put it on the back burner....* **Human activity is the cause of the climate problem.... We are getting dangerously close to the point of no return.** *Humanity has a choice:* **cooperate or perish.** *It is either a* **Climate Solidarity Pact** *– or a* **Collective Suicide Pact.** *(emphasis mine)*[130]

On September 20, 2023, Guterres repeated his warning. He zeroed in on the human tragedy of the climate elephant in the room. He said: "Humanity has opened the gates of hell. Horrendous heat is having horrendous effects. Distraught farmers [are] watching [their] crops carried away by floods. Sweltering temperatures [are] spawning disease[s]. And thousands [are] fleeing in fear as historic fires rage. Climate action is dwarfed by the scale of the challenge. If nothing changes, we are heading towards a 2.8°C temperature rise – towards a dangerous and unstable world."[131]

A *New York Times* reporter, Somini Sengupta, was startled by the apathy of the audience, made up primarily by world leaders. She described how frustration, distrust, and impatience dominated the climate talks. The air at the Egyptian resort was not refreshing. Too much anxiety and too much bad news diminished hope. "Big polluting countries [like China and the United States]," she said, "aren't cutting their emissions of planet-warming gases as quickly as they need to. The money that rich countries promised to help poor countries transition to clean energy hasn't

---

[130] UN Secretary General's remarks to High-Level persons of COP 27, Sharm el-Sheik, Egypt, November 7, 2022.

[131] UN Secretary General, Secretary-General opening remarks at the Climate Ambition Summit, UN, New York, September 20, 2023.

materialized… many of the splashy goals that countries and companies set for themselves at last year's conference [in Scotland] haven't been met, like curbing deforestation or phasing down coal."[132]

Before the climate summit in Egypt, on October 27, 2022, the UNEP put substance to the warnings of Guterres. "As growing climate change impacts are experienced across the globe, the message that greenhouse gas emissions must fall is unambiguous," the UNEP report said. UNEP *called for rapid transformation of societies. It said* that the international community was "falling far short of the [2015] Paris goals, with no credible pathway to 1.5°C in place. Only an urgent system-wide transformation can avoid climate disaster."[133]

In the November 2021 UN climate meeting, in Glasgow, Scotland, nations promised to "phase down" coal. Yet in late 2022, coal consumption was soaring. Coal is at the center for the production of steel, iron, cement, electricity and carbon dioxide.[134] Countries had promised to "phase out" subsidies to fossil fuels. They did not. The Organization for Economic Cooperation and Development reported in 2022 that the subsidies OECD countries, including America, spent on fossil fuels almost doubled in 2021 to a total of US$697.2 billions. Of that largesse, US$302.7 billion went to petroleum, US$166 billion for natural gas, US$19.2 billion for coal, and US$69.04 billion for oil price support. In 2022, the OECD support of fossil fuels reached 1.4 trillion.[135] The same situation happened to the promises 197 countries made for ending emissions of the potent Earth-warming methane gas and the destruction of the forests. However, the OECD subsidy estimate may be low. The International Monetary Fund estimated the total subsidies to fossil fuels in late 2022 topped US$7 trillion.[136]

Alas, Guterres is not getting his points across successfully. Unless we reverse the heavy dependence we have on fossil fuels, we are doomed. The support of the fossil fuel industry by world leaders must be exposed for the fraud it is. Guterres, once again, denounced the promises of world leaders for net-zero emissions. On November 8, 2022, he said: "Using

---

[132] Somini Sengupta, Hello from COP27, *New York Times*, November 7, 2022.

[133] UNEP, *Emissions Gap Report 2022: The Closing Window – Climate Crisis.*

[134] International Energy Agency, Coal 2022, December 16, 2022.

[135] OECD, Fossil fuel support, 2022.

[136] Stefan Anderson, Fossil Fuel Subsidies Hit Record $ 7 Trillion in 2022, *Health Policy Watch*, August 30, 2023.

bogus 'net-zero' pledges to cover up massive fossil fuel expansion is reprehensible. It is rank deception. This toxic cover-up could push our world over the climate cliff. The sham must end. Second, on credibility, full and rapid decarbonization this decade is the ultimate test."[137]

Guterres' denunciation of the fake promises of world leaders about net-zero emissions is dated to 2022 but mirrors his entire experience in climate negotiations. World leaders forgot the 1997 Kyoto Protocol and the 2015 Paris Agreement that obliged them to reduce greenhouse gas emissions.

As usual, during the UN climate conferences, young people protested the actions of world leaders, pretending they favored action against climate change while, simultaneously, funding fossil fuels. The banners of the environmentalists at Sharm el-Sheikh, Egypt, said, "Stop funding fossil fuels! Stop funding death!" Dipti Bhatnagar, a young woman from Mozambique representing the Friends of the Earth International, shouted: "Rich countries are out to grab the huge gas reserves, and people are being dispossessed of their land. One million people out of the 23 million [of Mozambique's] population are living in refugee camps because of gas. We say no to more gas finance. We won't let Africa burn."[138] Kausea Natano, Prime Minister of Tuvalu, a tiny island nation in the Pacific, proposed a treaty to stop the expansion of fossil fuels, treating them like weapons of mass destruction. He sent this letter to world leaders:

"Dear world leaders at the COP[27, Sharm el-Sheikh, Egypt],"

"Climate change is drowning the Pacific Islands."

"The world's addiction to oil, gas and coal threatens to swallow our lands under the warming seas – inch by inch. But we will not stand by as our home is wiped from the map! So we're uniting with a hundred Nobel laureates and thousands of scientists worldwide to urge world leaders to join the Fossil Fuel Non-Proliferation Treaty to manage a just transition away from fossil fuels. The time has come to make peace with the planet. To deliver [to] vulnerable nations the long overdue funding needed to cope with the loss and damage incurred from climate disasters and to make polluters pay. They say that one day, the oceans will

---

[137]UN Secretary-General's remarks at launch of High-Level Expert Group on Net-Zero Commitments. Sharm el-Sheikh, Egypt, November 8, 2022.
[138]UN News, November 9, 2022.

swallow the place we call home. But I promise you this: until that day comes, we will keep fighting. Because if we can save our islands, we can save the world."[139]

Sun and Earth

The courage and vision of this world leader is necessary to stop the fossil fuel producers. In his own quiet way, former Vice President Al Gore shouted, too. His Climate TRACE Coalition[140] is capable of revealing the hitherto secret emissions of carbon dioxide, methane, and other greenhouse gases warming the planet. On November 9, 2022, Guterres honored Gore at Sharm el-Sheikh, Egypt. He said:

Climate TRACE and its data show that because of underreporting of methane leaks, flaring, and other activities associated with oil and gas production, emissions are many times higher than previously reported. This should be a wake-up call for Governments and the financial sector, especially those that continue to invest in and underwrite fossil fuel pollution. The problem is even greater than we were led to believe and that means we must work even harder to accelerate the phas[ing] out of all fossil fuels.[141]

The potential revelations of Gore's climate initiative will probably confirm Guterres' fear that the danger of the unleashed climate catastrophe is much larger than we have speculated from incomplete – nay, misleading – data. It turns out satellites track down more than 72,000 carbon polluters on the planet. One of the largest is a steel plant in Zhangjiagang, China, belonging to the Shagang Group. This steel plant has been "churning out tens of millions of tons of steel a year – and immense quantities of planet-warming gases." Climate TRACE estimated this plant is the most carbon-polluting factory on the planet. Climate TRACE boasts it can monitor carbon pollution of entire countries as well as industries, and individual factories, cargo ships, animal farms, steel,

---

[139] COP 26, Sharm el-Sheikh, Egypt, November 8, 2022. https://unfccc.int/sites/default/files/resource/TUVALU_cop28cmp18cma5_HLS_ENG.pdf.
[140] The Climate Reality Project, The Game Changing Promise of Climate Change.
[141] UN Secretary-General, Secretary-General's remarks at launch of Al Gore's climate TRACE initiative, Sharm el-Sheikh, Egypt, November 9, 2022.

cement and fertilizer factories, including gas and oil extracting facilities. Overall, Climate TRACE is giving us a compendium of more than 72,612 polluters. "a hyperlocal atlas of the human activities that are altering the planet's chemistry."[142]

Gore has no doubt that the TRACE carbon project, which he supports, has the potential of bringing some good for humanity. I am confident it will. Gore has been an environmentalist and student of climate change for several decades. Collecting the data of polluters inevitably has the potential of becoming some kind of carbon surveillance state, worldwide, "one which, even independent of any global enforcement mechanism, promises to change some aspects of the conventional picture of climate change and what is causing it."[143] I probably agree with this logical assessment.

Antonio Guterres is right. Enough with treating civilization and Mother Earth like a toilet. The fossil fuel companies are pushing us toward a hellish future. We must ban them – before 2030. We certainly do not want a global temperature rise of 2.7°C. Take science seriously. Act now to defend the Earth and civilization.

Pope Francis and International Climate Politics

Like Guterres, the late Pope Francis has become another environmental hero telling the truth. In his October 4, 2023 letter Laudate Deum[144] (which translates as "Praise God"), Pope Francis was more straightforward and science-based in his sentiments regarding the climate change crisis than ever before. He addressed this 7,000-word letter to "all people of good will" and he defined the purpose of his apostolic exhortation to highlighting the political and technological complexity of the climate crisis. With the exception of a few sections at the end dealing with spiritual and Biblical sayings about the Earth, most of this treatise is a brief and admirable summary of climate science that explains the changing climate and the resultant global warming. But what caught my attention

---

[142] Raymond Zhong, Who's driving climate change? New data catalogs 72,000 polluters and counting, *New York Times*, November 9, 2022.

[143] David Wallace-Wells, The global carbon surveillance state is coming, *New York Times*, November 16, 2022.

[144] The Holy See, Apostolic Exhortation: *Laudate Deum of the Holy Father Francis: To all people of good will on the climate crisis*. https://www.vatican.va/content/francesco/en/apost_exhortations/documents/20231004-laudate-deum.html.

was his achievement in addressing science in a philosophical and political context, something that matters today, in the third decade of the twenty-first century.

In his letter, Pope Francis added his voice to those of the scientists who have been urging world leaders to act before it is too late. He rattled data on extreme heat waves, draught, floods, forest fires, the melting of Arctic ice and the thawing of permafrost. He said these man-made phenomena have already damaged life in the oceans – harm likely to last for centuries. He subtly criticized the global United Nations leadership on climate; for having an annual Conference of the Parties, during which "the accords have been poorly implemented, due to a lack of suitable mechanism for oversight, periodic reviews and penalties in cases of noncompliance." He also lamented that "international negotiations cannot make significant progress due to positions taken by countries which place their national interests above the global common good." He mentioned that all global climate summits since 2015 in Paris had showed the failure to do what was needed.

The Pope was probably anxious about the potential outcome or failure of the COP28, which took place in December 2023 in oil-rich Dubai. He said that "[to] say that there is nothing to hope for would be suicidal, for it would mean exposing all humanity, especially the poorest, to the worst impacts of climate change." Pope Francis expressed his anger with the failing international system that risk the health and survival of humanity and Mother Earth. "I have realized," he said at the beginning of his letter, "that our responses [to the climate crisis] have not been adequate, while **the world in which we live is collapsing and may be nearing the breaking point**." (Emphasis mine)

This language caused anxiety in America. The Pope, is "a climate alarmist, a techno-skeptic and a degrowther," said David Wallace-Wells, a young reporter for the *New York Times*, "sympathetic to activists… [He has] also emphatically endorsed the 'abandonment' of fossil fuels – outing himself as a 'keep it in the ground' guy as well."[145]

But the Pope speaks the truth, like Antonio Guterres. Nothing he said alarmed me. Are we to dismiss such a world celebrity defending the Earth because he speaks of "the ethical decadence of real power," including real power in America? It seems to me that Pope Francis wrote for

---

[145] David Wallace-Wells, The Pope's journey to climate outrage, *New York Times*, October 11, 2023.

all humanity. He was right in denouncing "the technocratic paradigm of power" encoded into Artificial Intelligence (AI), which he called a "monstrosity" feeding on itself. I agree with Pope Francis. My view is that AI is not intelligence but a weapon like a nuclear bomb. Pope Francis cited numbers showing the steep rise of global temperature. "In the last fifty years," he said, "the temperature has risen at an unprecedented speed, greater than any time over the past two thousand years. In this period, the trend was a warming of 0.15°C [Celsius] per decade, double that of the last 150 years. From 1850 on, the global temperature has risen by 1.1°C, with even greater impact on the polar regions. At this rate, it is possible that in just ten years we will reach the recommended maximum global ceiling of 1.5°C. This increase has not occurred on the Earth's [land] surface alone but also several kilometers higher in the atmosphere, on the surface of the oceans and even in their depths for hundreds of meters. Thus the acidification of the seas increased, and their oxygen levels were reduced. The glaciers are receding, the snow cover is diminishing, and the sea level is constantly rising."

Pope Francis understood that harming or killing life in the vast oceans may have such immense consequences that the Biblical Apocalypses flashed across his mind. "Certain apocalyptic diagnoses may well appear scarcely reasonable or insufficiently grounded," he said. "This should not lead us to ignore the real possibility that we are approaching a critical point. Small changes can cause greater ones, unforeseen and perhaps already irreversible, due to factors of inertia. This would end up precipitating a cascade of events having a snowball effect. In such cases, it is always too late, since no intervention will be able to halt a process once begun. There is no turning back."

Apocalypses or not, tipping points are travels of no return. The dinosaurs and countless species have been lost forever. Theoretical criticism of the continued use of fossil fuels triggering such cosmic havoc has its limits. It appears that the fossil fuel oligarchs have influenced scientists, professors and politicians to keep the people misinformed and misguided while they enjoy royal powers and luxuries.[146,147]

---

[146] David F. Noble, *The Religion of Technology* (New York: Alfred A. Knopf, 1997).

[147] David F. Noble, *America by Design* (New York: Oxford University Press, 1977).

Any Way Out of Our Climate Tragedy?

Is there a way out of such a deadly dilemma? Do we keep writing that increasing global temperatures are not good for us and Mother Earth? And what happens, as the situation is developing today, when, with the rare exception of the UN Chief, Antonio Guterres, the vast number of politicians are not doing enough to make a reversal for climate change possible? In *Laudate Deum*, Pope Francis voiced his concern that "the necessary transition towards clean energy sources such as wind and solar energy, and the abandonment of fossil fuels, is not progressing at the necessary speed. Consequently, whatever is being done risks being seen only as a ploy to distract attention." Do we go on tolerating inactions from our world leaders and minimal measures such as increasing the number of electric cars and increasing investments in solar and wind energy, while the petroleum, natural gas, and coal magnates keep excavating the planet?

Pope Leo XIV is following the climate ideas and advocacy of Pope Francis. He loves biodiversity and the Earth. His hope is that the people will save the Earth.[148] I believe that the Pope could take his knowledge and genuine goodwill a step further. He does have access to the fossil fuel magnates of the world. Perhaps if he uses his influence as the Pope, he could persuade them to make drastic changes to reverse climate change. These magnates have the resources to move away from fossil fuels quickly and replace them with renewable energy worldwide. Perhaps, Pope Leo could convince them to fund a World Environment Organization (WEO) to clean up the pollution of the oceans and land. In theory, the Pope could go down in history as the modern protector of Mother Earth.

Nuclear weapons must be dismantled and abolished as rapidly as fossil fuels. Agriculture cannot go on poisoning the land and the very food people eat. Billions of petroleum-powered cars and trucks must also be recycled. Electric trams, trains, subways, and buses should do the heavy lifting in moving people and goods around. Plastic nets scooping up fish in the oceans and seas cause enormous harm to fish and marine biodiversity and ecosystems.[149] Just like factory agriculture, factory fishing must be reformed or abandoned. A WEO inspired by Pope Leo XIV could

---

[148] Motoko Rich, Pope Leo calls for unity on climate at a divided moment, *New York Times*, October 1, 2025.

[149] Jacob Shea, Ghost fishing nets: Invisible killers in the oceans, *Earth Island Journal*, January 7, 2014.

function to protect biological diversity and ecosystems. At least, that is my dream, which Pope Leo XIV might spark for our badly divided world. Perhaps, environmentalists and scientists and citizens can lead a movement to decarbonize America and, with the collaboration of similar movements from other countries, to facilitate the transition of energy from fossil fuels to harmless solar and wind energy.

Despite my reservations about the military industrial complex, as long as President Trump is in office, the US will keep using fossil fuels. However, I urge his successor to order the Corp of Engineers of the military, in cooperation with other government agencies and civil authorities and businesses, to remove fossil fuels and replace them with the gifts of nature, the sun and wind. The President can follow the paradigmatic policies of Franklin Roosevelt and order American corporations to abandon their conventional energy business practices and invest their talent and money in the transformation of the country from the world's top fossil fuel user to the world's top user of green technologies. Even the military will have to shed its dependence on petroleum. The harvest of such an experience would be an acknowledgment of nature and science-based policies. We need to realize, however, that solar and wind energy will not satisfy all Americans or some other people on the planet. Geographic and climatological conditions may make energy from the Sun difficult or impossible. In such cases, the state can make efforts to obtain green energy from countries with sources of energy unaffected by clouds or other weather conditions.

If the Presidents of America begin to declare war against climate chaos and set the foundations and infrastructure for renewable energy, we should have a fighting chance. But for that to happen, they must reconsider the Cold and Hot War against Russia.[150] President Biden understood the danger. On February 10, 2022, he urged Americans in Ukraine to leave the country immediately. "We're dealing with one of the largest armies in the world," Biden said. "This is a very different situation, and things could go crazy quickly. That's a world war when Americans and Russians start shooting at one another. We're in a very different world than we've ever been in."[151,152]

---

[150] Eli Stokols, Biden vows to end German - Russian gas pipeline if Ukraine is invaded, *Los Angeles Times*, February 7, 2022.

[151] Aishvarya Kavi, U.S. says Russia could invade Ukraine at any time, *New York Times*, February 11, 2022.

[152] Anton Troianovski, Andrew E. Kromer, Lara Jakes, and Katie Rogers, In phone call, Biden warns Putin of severe costs of invading Ukraine, *New York Times*, February 12, 2022.

War Clouds Intensify Climate Danger

On February 22, 2022, Russia invaded Ukraine. This invasion was part of a war that started in 2014. But Europe and America moved rapidly into a frenzy of Cold War rhetoric, threats and even racism against Russia. It feels like world leaders had replaced reason and examination of why the war took place. Jeffrey Sachs said that America and its NATO allies conveniently forgot that American money and influence overthrew the pro-Russia leader of Ukraine, President Victor Yanukovych, in February 2014. This American interference in the politics of Ukraine provoked the "shooting war" in Ukraine, which started then, in February 2014, not in February 2022.[153] But American and European leaders keep talking of an unprovoked Russian invasion of Ukraine. They kept uttering threats against Russia and imposing sanctions, endless financial and trade sanctions.[154] And, as fast as they could, they exported guns to Ukrainians, urging them to fight to death.

Stop Burning Fossil Fuels

This is a time Americans and Russians, and the rest of humanity should be very close to one another, planning and executing painful policies to secure a livable future for our children. This means, above all, drastically reducing greenhouse gas emissions from burning fossil fuels. An April 2022 editorial in the *Los Angeles Times* spoke exactly like Guterres and Pope Francis: "Stop burning fossil fuels," the editorial said. "End the construction and operation of coal, oil, and natural gas infrastructure. Dramatically accelerate the switch to clean renewable energy. Cut greenhouse-gas emissions nearly in half by 2030 and get to net-zero by mid-century."[155]

By 2024, sales of electric vehicles "boomed." Generating electricity from the sun, wind and other nonpolluting sources "bogged down" primarily from logistical reasons.[156] But with Trump back to power, climate

---

[153] Jeffrey Sachs, The war in Ukraine was provoked – and why that matters to achieve peace, *Common Dreams*, May 23, 2023.

[154] Matina Stevis-Gridneff, European diplomats slowly close in on new Sanctions, adding limitations, *New York Times*, April 7, 2022.

[155] Climate clock is counting down, *Los Angeles Times*, April 5, 2022.

[156] Brad Plumer, Here's Where Biden's Climate Law Is Working and Where It's Falling Short, *New York Times*, Feb. 21, 2024.

"faded from the national agenda." Even the words climate change disappeared among bureaucrats of the federal government.[157] Unfortunately, Senator Joe Manchin of West Virginia forced the federal government to continue to offer public lands and waters to petroleum, coal, and gas companies for the extraction of fossil fuels. "This is a climate suicide pact. It's self-defeating to handcuff renewable energy development to massive new oil and gas extraction. The new leasing required in this bill will fan the flames of the climate disasters torching our country, and it's a slap in the face to the communities fighting to protect themselves from filthy fossil fuels," said Brett Hartl, government affairs director at the Center for Biological Diversity,[158] one of the most important environmental organizations in the country. "The bill would... revive offshore oil and gas lease sales in the Gulf of Mexico and Alaska," said Hartl. "The fossil fuel industry got its vile tribute, extracted with cruel precision by a profiteering coal baron... A reported side deal between Sen. Manchin and Democratic leadership would weaken bedrock environmental laws to give faster fossil fuel project approvals and expedite the fracked gas Mountain Valley Pipeline in Appalachia. Fossil fuel development is the root cause of the deadly fires, floods and heatwaves ripping across the country...The heat domes, deadly fires and ravaging floods punishing our country won't wait for the next anguished compromise. President Biden must act to end the fossil fuels destroying lives and our planet before it's too late."

True, the Manchin compromise might water down the benefits of spending lots of money for fighting climate change. There is hope, however, that the Center for Biological Diversity and the rest of the environmental community will fight any new fossil fuel leases the Department of Interior considers for approval. Yet spending US$369 billion for climate

---

[157]Brad Plumer and Lisa Friedman, Don't Mention Climate: Now, Clean Energy Is All About the Money, *New York Times*, May 21, 2025.

[158]H. R. 5376 – Inflation Reduction Act of 2022 or Public Law No: 117–169. US Senate Democrats, *The Inflation Reduction Act of 2022*. Press Release, Manchin poison pills buried in inflation reduction act will destroy livable climate, July 28, 2022: Climate bill heads to Biden for signature, forecasts fall fossil fuel fight, Center for Biological Diversity, August 12, 2022.

protection is a big deal. It is the strongest climate action by the United States to fight climate change.[159,160]

An additional important development of largely symbolic but real significance took place at the UN. The UN General Assembly passed a resolution declaring that a clean, healthy, and sustainable environment is a universal human right. Antonio Guterres urged states to implement the resolution and make a healthy environment available to all everywhere.[161]

Al Gore warned that fossil fuel billionaires intend to continue corrupting every rank and file politician and government official, including the bureaucrats of the UN, in an interview with the *New York Times* in September 2023.[162] This is especially evident with the choice of "a top oil executive, Sultan al-Jaber of the United Arab Emirates," administering the December 2023 Climate Summit in Dubai. "That's just, like, taking the disguise off," Gore said. "They have captured control of the political and policymaking process in too many countries and too many regional governments, and they've reached out to try to capture the U.N. process... Fossil fuel industries... have portrayed themselves as the source of trusted advice that we need to solve this crisis. But they are responding to powerful incentives to keep digging and drilling and pumping up the fossilized remains of dead animals and plants and burning them in ways that use the atmosphere as an open sewer, threatening the future of humanity." Gore described the Big Oil corruption in the same interview with a *New York Times* reporter.

The *New York Times* also talked to a petroleum billionaire, Michael Bloomberg who defended the right of the petroleum men to keep making money from selling petroleum to people. He supported al-Jaber leading the talks at the Conference of the Parties #28 in November–December 2023. "We are not going to get away from using oil for the next 10 or 15 years and we are not going to say everybody that has a gas-guzzling car can't drive it anymore and they will have to start walking today," he

---

[159] Emily Cochrane, House passes sweeping climate, tax and health care package, *New York Times*, August 12, 2022.

[160] German Lopez, Ditching fossil fuels, *New York Times*, August 12, 2022.

[161] UN General Assembly declares access to clean and healthy environment a Universal human right, *UN News*, July 28, 2022.

[162] Vivian Giang, Al Gore says fossil fuel industry seek to 'capture' climate talks, *New York Times*, September 21, 2023.

said. "Big oil is part of the problem. They are also part of the solution."[163] I disagree. Big oil executives are not part of the solution of a tragedy they created and supported for decades. In the interview with the *New York Times*, Gore politely suggested these guys remove themselves from their fraud and let the rest of Americans build a new energy world. He said fossil fuel companies "don't disclose their emissions. They don't have any phase-out plan. They're not committed to a real net zero pathway. They're greenwashing. They're performing anti-climate plotting."[164]

I would go a step further than Gore. Tax and fine fossil fuel companies for the colossal damages they have inflicted on us and the planet. And use those trillions to speed up the transition to renewable energy. Of course, this is a pie in the sky, especially now that Americans have re-elected Trump to the presidency. The man denies climate change. He appointed two oil and gas corporate executives, Doug Burgum and Chris Wright, to head the Departments of Interior and Energy, respectively. Burgum, governor of North Dakota, is an oil man and Wright is also a petroleum franking businessman. They promise big trouble for America's struggle to switch from fossil fuels to alternative clean energy, such as solar and wind energy. Manish Bapna, president of the Natural Resources Defense Council, called these appointments "deeply troubling," but not destructive. "The clean energy transition in the United States," he said, "is unstoppable."[165] David Gelles,[166] and his colleagues at the *New York Times*, are also optimistic about the energy transition in America. Bapna criticised Trump's appointees in the Energy and Interior Departments. He said they are unqualified to support the public interest and the health of nature. They are billionaires from oil fracking and oil business that will continue their corrupt practices in office. Bapna expressed great concern for the safety and health of Americans under these Trump officials. They speak of a "profound shift" taking place from the transformation and replacement of the fossil fuels-powered technologies and industries to clean alternatives. "Wind and solar power," they say, "are breaking records, and

---

[163] Manuela Andreoni, The fault lines at climate week, *New York Times*, September 21, 2023.

[164] David Gelles, Al gore on extreme heat and the fight against fossil fuels, *New York Times*, July 18, 2023.

[165] Democracy Now, Trump picks climate-denying oil and gas magnate as energy secretary. He once drank fracking fluid on live TV, November 18, 2024.

[166] David Gelles *et al.,* The clean energy future is arriving faster than you think, *New York Times*, August 17, 2023.

renewables are now expected to overtake coal by 2025 as the world's largest source of electricity. Automakers have made electric vehicles central to their business strategies and are openly talking about an expiration date on the internal combustion engine... governments around the world are pouring trillions of dollars into clean energy to cut the carbon pollution that is broiling the planet." Ironically, the reporters also said that despite these signs of a hopeful future, fossil fuels still dominate energy "at home and abroad." The scale of rebuilding for clean energy and removing fossil fuel is "mind-boggling." Ultimately, it is not clear if the US and other countries will stop greenhouse gas emissions to avoid catastrophe. He keeps calling climate change a hoax and a con job, according to a report about taxing shipping emissions: Trump "bullied" states to abandon their green energy plans. He uses "trade talks and tariff threats and verbal dressing-downs."[167]

Geocentric Cosmos

This looming reality (of fear, danger, anger, massive ecocide and hope) is another reason why the ancient Greeks are our mentors. They worshipped the Sun god Helios for millennia. Did they know something about the Cosmos that, in our hubris, we ignore? That the sun, for all intents and purposes, is nearly forever? NASA says the Sun is 4.6 billion years old, with another 5 billion years left. However, 5 billion years is an immense timeline. Humans rarely live more than 90, much less, 100 years. The sun is life-giving and light-giving. The Greeks called the sun "Heelios/Helios" because Helios means "the gathering of people observing the rise and setting" of this magnificent star. In the first chapter, I discussed the contribution of the Greeks to agriculture and civilization. The Greeks put the Earth (Gaia) at the center of the universe. We describe that cosmological design as the geocentric universe. This shows the immense respect Greeks had for the Earth as a living being. The *Homeric Hymn to Gaia* describes the Earth as mother of the gods and wife of the heavens; the very ancient Mother of All, which nourishes every single plant and animal.[168] Plato, too, thought the Earth

---

[167] Jennifer A. Dlouhy and Akshat Rathi, How Trump pressures the world into burning more oil and gas, *Los Angeles Times*, October 31, 2025.
[168] *The Homeric Hymns* 30: To Earth, Mother of All.

was the first and oldest of the gods, our foster-mother and the guardian and maker of day and night.[169]

In the third century BCE, another natural philosopher, Aristarchus of Samos, put the Sun at the center of the Cosmos. Aristarchus' heliocentric cosmology best explains how the universe works. It's our cosmology. However, in my opinion, modern men and women and most scientists look at the Earth as a mine for wood, fish, fossil fuels, metals, minerals, and food, what economists describe as "resources." With such a mentality, it explains the hunting and killing of terrestrial and marine wildlife and the ruthless treatment of our planet: dumping greenhouse gas emissions in the atmosphere,[170] clearcutting forests, exploiting and polluting the seas,[171,172,173] bleaching the corals,[174] and transforming ancient, gentle and ecological practices of growing food into mechanical factories that poison the land and the very food people eat.[175]

The next chapter expands on the arguments of Chapters 5 and 6, illustrating the dangers of animal farms. These animal factories are, in my view, major sources of disease, pollution, and global warming gases. Along with potentially accidental releases of pandemic-capable viruses from biological war labs, industrialized agriculture, fueled by pesticides and animal farms, appear to be undoing centuries-old of sustainable farming and rural life.

---

[169] Plato, *Timaeus*, 40.

[170] Naomi Oreskes and Erik M. Conway, *Merchants of Doubt: How a Handful of Scientists Obscured the Truth on Issues from Tobacco Smoke to Global Warming* (New York: Bloomsbury Press, 2010).

[171] Pat Costner and Joe Thornton, *We All Live Downstream: The Mississippi River and the National Toxics Crisis* (Washington, DC: Greenpeace, 1989).

[172] Marc Reisner, *Cadillac Desert: The American West and Its Disappearing Water* (London: Penguin Publishing Group, 1993). pp. 452–514.

[173] Beth Millemann, The dirty deeps, *Sierra*, **75** (3), May/June 1990, pp. 30–32.

[174] Catrin Einhorn, The widest-ever global coral crisis will hit within weeks, scientists say, *New York Times*, April 15, 2024.

[175] Vallianatos, *Fear in the Countryside* (1976). *Poison Spring* (2014).

7 Big Farms: Raising Food or Pestilence?

Imaginary face of pestilence.

Photo Courtesy of Nathaniel St. Clair.

Animal Slaughter and Disease

There are 1365 chicken growers and 10 processing plants for the millions of chicken slaughtered in the Delmarva region of Maryland, Delaware, and Virginia. But the explosion of the coronavirus pandemic in 2020 threw more than uncertainty into the gears of these large animal factories. It brought to light their extreme vulnerabilities: their difficulties to adjust to a reduced workforce. The machines had to have so many workers to process so many chicken per hour. Workers had to debone meat at a predetermined rate. They had to themselves almost become machines – or at least part of the unthinking machine. They dress in plastic uniforms and work shoulder to shoulder. It was alleged that workers wear diapers and were denied bathroom breaks because the machine was programmed to dump dead chickens to the moving line leading to the knives-yielding workers. Throughout this cruel, hard, and difficult daily experience, the workers are dripping with blood, pieces of animal flesh and excrement.[1-3]

This is nothing new. A few years ago, Ellen Silbergeld, professor emerita of toxicology at Johns Hopkins University, wanted to find out how workers in meat slaughtering and processing plants saw food safety in relation to their own safety. She found a book that revealed the horrors of animal slaughter and food safety at meat processing in Chicago's slaughterhouses in the early twentieth century. That book was *The Jungle* by the investigative journalist Upton Sinclair.[4] The book was serialized in 1904–1905 and published in 1906. She gave *The Jungle* to Keith Ludlam, president of the meat processing workers in Tar Heel, North Carolina. Ludlam read the book and asked Silbergeld when the book was written. She told him. He furiously said: "Nineteen-oh-four? Nineteen-fucking-oh-four? Nothing has changed! Nothing has changed at all!"[5,6]

---

[1] *Lives on the Line: The High Human Cost of Chicken* (Oxfam America, 2015). https://www.oxfamamerica.org/livesontheline/.

[2] *No Relief: Denial of Bathroom Breaks in the Poultry Industry* (Oxfam America, 2016).

[3] Christine Hauser, Nearly 2 million chicken killed as poultry workers are sidelined, *New York Times*, April 28, 2020.

[4] Upton Sinclai, *The Jungle* (New York: Doubleday, Page and Company, 1906).

[5] Bret McCabe, Ellen Silbergeld explores the dangers of Industrial meat production, *Johns Hopkins Magazine*, Fall 2016.

[6] Ellen Silbergeld, *Chickenizing Farms and Food: How Industrial Meat Production Endangers Workers, Animals, and Consumers* (Baltimore: Johns Hopkins University Press, 2016), pp. 1–2.

Silbergeld was probably as astonished as I am. An entire century without any improvements in work conditions! How did that happen? What did the scientific community do? I suppose that those hundreds of universities, scientific academies, environmental organizations were probably in deep sleep.

In 1906, however, Americans were outraged after reading the tragic and criminal revelations made by Upton Sinclair in his book. The result was the 1906 Pure Food and Drug Act and the founding of the US Food and Drug Administration (FDA). They sounded like good solutions – in theory. In reality, they did not touch the meat and slaughter monopolies.[7] More legislation passed after WWII to regulate the safety of food and food production workers: the Centers for Disease Control and Prevention (CDC), 1946; the US Environmental Protection Agency (EPA), 1970; and the Occupational Safety and Health Administration (OSHA), 1971. However, these wonderful laws and extensive bureaucracies, including the US Department of Agriculture (USDA), which was founded in the nineteenth century, have not been able to assure us that our food is safe to eat, nor do they guarantee that agricultural and meat processing workers labor under safe conditions. No state or federal regulation touches on animal farms.[8]

And when a pandemic strikes, as it did in 2020, animal butchers tend to come down with diseases from the animals they killed. They then pass the diseases on to each other, their families, and the rest of society. The other factor was that these diseases weaken their immune systems, predisposing them to much larger and deadlier diseases. In 2020, the CDC warned people with any "underlying medical conditions" that they were vulnerable to the coronavirus pandemic.[9] The warning was correct. Some of these people with underlying medical conditions did catch the coronavirus.

When the infected workers stayed home in 2020, their absence disrupted the chain of slaughter and meat processing. The owners of the chickens, hogs and cattle, slaughtered and alive, had the opportunity to rethink their mechanical monsters, scaling them down, or scrapping them altogether. This did happen to some of these mechanical farms, but

---

[7] Eric Schlosser, *Fast Food Nation*, pp. 204–207.

[8] Jon Devone, Valerie Baron, D. Lee Miller and Gregory Muren, CAFOs: What we don't know is hurting us, Natural Resources Defense Council, September 23, 2019.

[9] US Centers for Disease Control and Prevention, People of any age with underlying medical conditions, July 17, 2020.

very slowly. Tyler Whitley knows something about the difficulties of poultry farmers to abandon animal farms. He directs an effort to help farmers who have decided to get out of animal farms to find another way of life, usually in organic farming. He calls his project "Transfarmation." He pointed out that 20% of those managing chicken farms live below the poverty line. He told a reporter that "I came to view factory farming as a cancer on rural America... I hated how it robbed people of their humanity and reduced them down to a number, to a widget to a cog."[10] Witley is funded by the non-profit organization, Mercy for Animals.

However, it would appear that animal farms dominate rural America. For example, the reporter Michael Pollan said that chickens, cattle, and hogs are marvels of "brutal efficiency, bred to produce at warp speed when given the right food and pharmaceuticals." The same thing, Pollan argued, happens to the factories killing and processing chickens, cattle, and pigs: their fuel is technically an unsustainable economic efficiency.[11] In the midst of the COVID-19 pandemic, slaughter factory owners defended their practices and continued with their businesses. With insufficient workers to process the animals they kill, they merely depopulated their chickens, killing and burying millions[12] without batting an eyelid. They also turned to President Trump to maintain the continuous operations of their meat factories. Trump dully "ordered" the meat factories to remain open because, in his mind, these factories were an integral part of America's "critical" infrastructure.[13]

Crippled Youth

That critical infrastructure did not include the harm of poverty, much less the harms of eating fast food. On February 5, 2012, the Secretary of the Navy, Ray Mabus, said to Brian Lamb of C-Span TV that three out of four young Americans aged 17 to 24 are disqualified to serve the armed forces because of health problems such as obesity, having a criminal record, and/or failure to finish high school. Ray Mabus, former governor of

---

[10]Cara Buckley, Life [of CAFO Farmers] After factory farming: The longer they're out, the happier they are, *New York Times*, August 29, 2024.

[11]Michael Pollan, The sickness in our food supply, *The New York Review*, June 11, 2020.

[12]Christine Hauser (2020), *Op. cit.*

[13]Taylor Telford *et al.,* Trump orders meat plants to stay open in pandemic, *Washington Post*, April 28, 2020.

Mississippi, shared the unfortunate truth during the interview: that 75% of young Americans are in such a bad state of being, either having bad health, being illiterate or having a criminal past. The military rejects their application for national service as they are deemed unqualified to serve.[14] He was convinced that something must be done to fix these endemic issues so that America can continue to have a robust military. I was shocked. I had not realized how impoverished America had become. Mabus was right to say that something must be done to secure the country's future by improving its key foundations: the health and prosperity of the population, in particular, the country's young people.

The military recognized the connection between obesity and national defense. More than 100 retired generals and admirals signed off on a report in which they urged Congress to "get the junk food out of our schools."[15] These senior military officials serving the country during the Obama administration expressed alarm that an estimated 26 million young Americans are either obese, poorly educated or have a criminal record.

On the surface, there is no obvious connection between big animal farms and obese young men and women flunking their military entrance exam. But these two events tell a story of a society in deep distress, with rising poverty and gross inequality; where government officials seemingly ignore the fate of millions of young Americans eating the wrong kind of food.

The pandemic has weakened Big Ag. The Food and Environment Reporting Network (FERN) painted a bleak picture: as of July 20, 2020, the pandemic was all over agribusiness facilities: about 504 meat packing and food processing factories (370 meat packing and 134 food processing) and 71 farms were afflicted by the coronavirus disease. This translated to some 45,807 coronavirus-infected workers (36,896 from meat packing and 4376 from food processing factories, including 4535 farm workers). About 188 workers died (168 from meat packing, and 14 from food processing plants and 6 from farms).[16]

---

[14]Evaggelos Vallianatos, Q and A, C-Span, February 5, 2012. https://www.c-span.org/video/?303887-1/qa-ray-mabus.

[15]Too fat to fight: Report says school lunches a threat to national security, *PBS News*, April 21, 2010.

[16]Leah Douglas, Mapping Covid-19 outbreaks in the food system, *Food and Environment Reporting Network*, April 22/July 20, 2020.

Chaos and Fear in the Countryside during the COVID Pandemic

In a documentary that aired on July 21, 2020 with the telling title, *"Covid's Hidden Toll,"* the PBS FRONTLINE program looked into the abuse of powerless meat and farm workers. The findings are not surprising. They mirrored the findings that I have already mentioned from my own research. It was suggested that slaughterhouse owners and large farmers value profit over the safety of workers in their facilities. By July 2020, close to 35,000 to 40,000 meat workers were probably sick with the virus as estimated by FERN. In California, about 93% of the farmworkers were Latinos. These agricultural workers lived in crowded housing, and were transported to job sites in packed vans. They were "bearing the brunt" of the pandemic since they had little access to healthcare, including testing and personal protective equipment. According to advocates for these agricultural employees, they had been treated as disposable; expected to risk their lives in heat waves and work in pesticide-laden fields. Naturally, they would get sick with the coronavirus.[17] The pandemic troubles that afflicted the impoverished farm workers, especially in the Central Valley of California, have been plowing them under.[18]

My personal opinion is that most people do not know where their food comes from: who produce the crops and how those crops become the food they buy at the grocery stores or supermarkets. Max Cuevas, a medical doctor managing clinics for farm workers in the Salinas Valley, California, was very unhappy. "I think the average American has no concept of how food reaches our table," he said in an interview with FRONTLINE.[19] "We don't know how meat is processed. We have no idea where lettuce comes from. We have no idea how it's harvested. I think there's a huge disconnect with those of us who have sheltered in place not understanding how those people [harvesting our food] work and how much they have to work to make a living and to make it profitable for the company that they're working for." Cuevas touched on the civilization shift in America, the massive post-WWII forced exodus of small family farmers from rural

---

[17] Anita Chabria, Farmworkers were hit hard, *Los Angeles Times*, July 25, 2020.

[18] Rong-Gong Lin II, Melody Gutierrez and Anita Chabria, Plowed under once again, *Los Angeles Times*, July 28, 2020.

[19] Rosa Turan and Nick Roberts, Farmworkers are among those at highest risk for COVID-19, *FRONTLINE*, July 21, 2020.

America and the takeover of all that fertile land by large farmers and corporations who hire armies of foreign workers to grow and harvest America's food. Now, seventy-five years later, the children of all those millions of former farmers are city people who, in fact, know little, if anything about growing food. Cuevas is angry with the injustice of the government and corporations towards farm workers. He feels the pain of the undocumented farm workers from Mexico who work so hard and live precarious lives in the sick farms of California.

The interview of Cuevas is complemented with quotations from the conversations of farm workers from the transcript of the FRONTLINE documentary cited in the footnotes.[20] I want readers to get a feeling for the harsh reality governing our broken and diseased agriculture. The dry and tight dialogues speak volumes about the environment of fear and disease hovering over these workers and the reporters interviewing them. The workers want to go away, but they have no place to go to. They want to speak the truth, but fear clouds their minds, putting a brake to their thoughts and tongues.

The FRONTLINE documentary, "Covid's Hidden Toll," speaks of more than 90,000 farm workers in Monterey County, California. These men and women plant, harvest and package our food. They live in "extremely crowded dwellings," according to the farm workers interviewed. There is no way they could have avoided the virus. The documentary's interviews also highlighted what was at stake during the pandemic. Workers were risking their lives for puny salaries – and insults. Yet the workers were, and still are, the gears for the profits of food corporations. In my opinion, these capitalist institutions are heartless, shameless, and ruthless. Farm workers pay a very heavy price for earning a living in a dangerous environment. They sleep in crowded housing.[21] The FRONTLINE documentary allows us to peer behind the curtains separating us from those feeding us: the farm workers. It throws light on the politics and practices of a slaughterhouse, meat processing plant, and vegetable farm in the Central Valley of California:

FRONTLINE reporter: DAFFODIL ALTAN: Central Valley Meat employs around 700 people at its plant in Kings County. The company

---

[20] PBS, FRONTLINE, Covid's Hidden Toll, July 21, 2020.

[21] Melissa Gomez, We can't prevent it: Farmworkers paid low wages fear coronavirus spread in crowded housing, *Los Angeles Times*, June 9, 2020.

has a history of violating health and safety codes, and has been cited for animal abuses... In the last decade it's had two beef recalls and been shut down three times. When the virus appeared at the plant in April [2020], workers told us that at first the company did nothing to protect them... Central Valley Meat ... has publicly denied threatening to fire workers or punishing them for being out sick. In late April, the company sent a note to employees comparing the outbreak to a normal flu season and saying that "the coronavirus is not some cloud floating around waiting to infect someone" and assured employees that nightly cleaning was killing any potential virus residue... I reached out to union leader Mark Lauritsen, who'd been monitoring outbreaks in plants around the country.

MARK LAURITSEN: Nobody knew exactly what they were dealing with... inside plants, it was chaos, it was fear, it was anxiety... Just in meatpacking alone in the United States, over 14,000 of our members have been exposed [to] or contracted COVID-19 because of their proximity to [each other]... if you look across the entire industry, you're probably looking at a number that's substantially higher than that. And when you have 14,000 of our members that are exposed and sick, that's a tremendous stress on the efficiency of the whole food supply chain in this country... if we want to protect our food supply chain in this country, let's protect those workers.

FRONTLINE reporter Altan concluded that "at least 35,000 meatpacking workers have been infected, with more than 100 deaths... [with] workers feeling pressured by their circumstances and by their employers."

Meatpacking factories in California were demanding their workers to put their lives on the meat cutting line. We learn from the FRONTLINE documentary that since April 2020, more than 53,000 food workers from all over the country have fallen ill from the virus. More than 200 of those ill workers died. The majority of the sick and dead workers made a living in the meatpacking industry. Meat workers often got ill and risk death while earning minimum wages. One of those Foster Farms workers, Martha Vera, worked for 24 years. She mourned for her husband who worked for Foster Farms as a truck driver and died from the Covid-19 virus. "What does this company really want?" she said, crying.

"How many more people do they think should die for them to do something... to protect [their] workers? How many more?"[22]

Thousands of meat processing workers came down with COVID-19. Others died from unsafe conditions at the very factories in which they worked. For example, six workers died from a leaking liquid nitrogen line in a poultry factory in Gainesville, Georgia, the so-called poultry capital of the world. Liquid nitrogen may evaporate, becoming deadly gas sucking air and oxygen, thus killing people by asphyxiation.[23]

The Dangers of Animal Farms

I think it is worth repeating that the owners of meat and vegetable factories, probably including the large farms growing organic fruits and vegetables, are more interested in maintaining their profits than protecting the lives and health of farmworkers. They buy that power with money. From 2019 to 2023, agribusiness corporations spent more than US$500 million lobbying Congress.[24] Lobbying is an influence on and the outright funding of the election and re-election of politicians agreeing with the agenda of large farm companies. These Congressional politicians then pass legislation that favor the interests of those who lobbied them. The government also reciprocates with deregulation and other measures favoring the lobbyists and the corporations they represent.

According to the Editorial Board of *The New York* Times,[25] Trump's (then) Secretary of the Department of Labor, Eugene Scalia, told OSHA to leave meat packing corporations alone. So, naturally, Trump's OSHA "has been asleep at the wheel." It issued recommendations to meat packers, not orders. When it fined irresponsible companies, the fines were tiny. This tells companies they have nothing to fear. Without enforcement, laws

---

[22] Jie Jenny Zou, Andrea Castillo and Erin B. Logan, Meatpacking was already risky. Then came the virus, *Los Angeles Times*, September 8, 2020.

[23] Richard Fausset and Michael Levenson, 6 die after liquid nitrogen leak at Georgia poultry Plant, *New York Times*, January 28, 2021.

[24] Omanjama Goswami and Karen Stillerman, *Cultivating Control: Corporate Lobbying on the Food and Farm Bill* (Union of Concerned Scientists, May 13, 2024).

[25] Editorial, Under trump, OSHA's COVID-19 response is failing workers, *New York Times*, September 14, 2020.

become invisible and companies don't see why they should "protect their employees and their communities."

Was the Joe Biden administration any different in protecting meat packing workers and, more fundamentally, in altering the foundations of inequality and danger so pervasive in the entire US agriculture?

In late February 2021, three family farmers[26] appealed to President Joe Biden to change America's farm priorities. They accused corporate meat packers of exposing their workers to unsafe conditions so that they could export more meat. "Lobbyists for big meat companies," they said, "have successfully weakened enforcement of worker safety, antitrust, and competition rules, and environmental and food safety protections... multinational meat companies acted like they always do: increasing profits while independent family farm livestock producers are paid less, workers are put at risk, and consumers pay more."

If meat packers treat workers with such contempt, imagine how they treat the animals. One needs to visit an animal farm to see the brutality they exercise and inflict on hundreds of millions of animals. Animal farms put cattle, pigs, chickens, and turkeys by the hundreds and thousands and millions very close to each other in confined spaces. The result is pestilence. A brief history explains why.

On December 1997, US Senator Tom Harkin (from Iowa) released a report[27] documenting the staggering amounts of waste coming out of those animal factories of livestock and poultry production in the United States. In 1997, the total animal waste was 1.37 billion tons of solid manure – 1,229,190,000 tons from cattle; 116,652,300 tons from hogs; 14,394,000 tons from chickens; and 5,425,000 tons from turkeys. This is as if every man, woman, and child in the country produced 5 tons of waste in 1997. The animals of giant agriculture in 1997 generated 130 times more manure than the human waste of all the people of the United States. What to do with such a mountain of manure when, for instance, outside of Washington, DC, in the Delmarva Peninsula, 600 million chickens produce 3.2 billion pounds of raw waste every year – releasing as much

---

[26] Bonnie Haugen, Darvin Bentlage and Barb Kalbach, President Biden, who controls our food system matters, *The Nation*, February 23, 2021.

[27] Senate Minority Staff Report for Senate Committee on Agriculture, Nutrition, and Forestry "Animal Waste Pollution in America: An Emerging National Problem, Environmental Risks of Livestock and Poultry Production," December 1997. https://hdl.handle.net/20.500.14300/130.

nitrogen in the environment as a city of nearly 500,000 people? A 50,000-acre swine factory in southwest Utah was designed to "produce" 2.5 million hogs every year. It was a project of the Virginia company Smithfield Farms, which sold it to a Hong Kong meat processor named Shuanghui International.[28] The potential waste per year of this Utah hog factory could be larger than the sewage of the city of Los Angeles, California. It is hardly surprising that staggering amounts of manure, concentrated at places where they could not be used as fertilizer, would cause the pollution of rivers and ground water, and other severe environmental problems. The Mississippi River, for example, drains a huge swath of the country; bringing animal waste, and fertilizer "nutrients" from farm run-off to the Gulf of Mexico/Gulf of America, resulting in thousands of square miles of the Gulf's water being almost devoid of oxygen – that is, it is dead. In 2020, the dead zone measured between 6700 square miles, the size of Connecticut and Delaware combined, and 7769 square miles, the size of New Hampshire.[29,30]

In August 2000, the Waterkeeper Alliance, a civil society organization from New York and several North Carolina environmental groups, filed a legal complaint[31] at Wake County, North Carolina, against Smithfield Foods and its subsidiaries – Carroll's Foods, Brown's of Carolina, and Murphy Farms. The complaint said that Smithfield Foods (and its dependent hog companies) have been poisoning the Neuse, Cape Fear, and New River of North Carolina, the basins of those rivers, and the land adjacent to the hog factories and slaughterhouse of the Smithfield Packing Company. Smithfield Foods, the suit charged, had been maintaining open hog cesspools full of untreated hog feces and urine and other wastes from pigs. This waste contaminated surface and ground water and aquifers as well as poisoned the air with ammonia from the nitrogen in the hog cesspools. The suit alleged that the owners of the hog factories willfully kept

---

[28] Dawn House, Utah hog farm part of $ 7.1 billion Chinese deal, *The Salt Lake Tribune*, July 12, 2013.

[29] Pat Costner and Joe Thornton, We All Live Downstream: The Mississippi River and the National Toxics Crisis (Greenpeace USA, December 1989), pp. 1–4.

[30] John Schwartz, Gulf of Mexico dead zone will be large this summer, *New York Times*, June 3, 2020.

[31] Twiggs, Abrams, Strickland and Trehy, Amended Complaint, State of North Carolina, Wake County, Neuse River Foundation, the Water Keeper Alliance v. Smithfield Foods, Inc., August 2000, paragraphs, pp. 1–169.

the hog cesspools overflowing with pig waste and did nothing to prevent the poisoning of the environment from their business; and that they also knew of the detrimental effects their hog farms were having on the people near their pigs. During September–October 1999, Smithfield Food's hog lagoons overflowed into the rivers of North Carolina, causing massive contamination of both surface and ground water. Smithfield Foods illegally sprayed untreated pig waste onto land that was not fit and large enough to absorb the waste. The company also sprayed huge amounts of untreated hog waste into the creeks and rivers of North Carolina. Among other deleterious substances, like copper and zinc, all this pig waste was equivalent to dumping more than 29,000 metric tons of phosphorous every year into the rivers and creeks of North Carolina.

According to the complaint, Smithfield buried thousands of dead and drowned hogs in its own sprayed fields. The result has been the massive contamination of water, land, and air. The worst poisoning of the water was in the river systems of the Neuse, Cape Fear, and New River. To look at it in a different light, Smithfield's volume of hog waste reaching the rivers of North Carolina is larger than that of the human waste of the entire population of North Carolina and New York state. In addition, pig waste is loaded with poisons, including heavy metals,[32] which poisoned the waters of North Carolina. Smithfield has caused devastating environmental damage to the rivers, creeks and natural environment of North Carolina. "Smithfield," the Complaint said, "has been poisoning the water, land, and air of North Carolina in a manner that is negligent, wanton, callous, and reckless. It operates with a disregard of the rights and safety of the public. Its business practices are abusive." And its conduct, as described in the suit, is "deceptive, immoral, unethical, oppressive, unscrupulous, and substantially injurious." In 2018, the Court sided with the Waterkeeper Alliance.[33]

Animal farms harm more than just North Carolina. Chris Jones, a water engineering professor at the University of Iowa, is well aware of the calamities of animal farms in Iowa. But he cannot bite the hand that feeds him. Wisely, he sought protection behind numbers. He crunched statistics in order to illustrate the impact of hogs on the environment and public

---

[32] Wenchong Lan *et al.*, Effects of application of pig manure on the accumulation of heavy metals in rice, *Plants*, January 14, 2022.

[33] Ellen Simon, Jury Finds for Smithfield Neighbors, December 12, 2018. https://water keeper.org/news/jury-finds-for-smithfield-neighbors/.

health of Iowa. His Blog[34] allows us to take a peek on the swine kingdom of this agricultural state. While the population of Iowa is only about 3 million people, the volume of animal waste alone makes Iowa a giant on the Earth. "Compared to humans, a 'feeder pig,' says Jones, 'excretes' 3 times more nitrogen, 5 times as much phosphorus, and 3.5 times more total solids than a human being." A hog at birth weighs about 3 pounds. In 6 months, that hog weighs 250 pounds. At that moment, the hog is sent to a slaughter factory. Jones figured Iowa in 2019 had "around 20–24 million hogs; 250,000 dairy cattle; 1.8 million beef cattle; 80 million laying chickens; and 4.7 million turkeys." He calculated the waste of these animals would be the equivalent of a much larger human population than the 3 million inhabiting Iowa at the time. Namely, the hogs were the equivalent of 83.7 million people, broken down as follows: "Dairy cattle: 8.6 million people; Beef cattle: 25 million people; Laying chickens: 15 million people; and Turkeys: 900,000 people." In short, these animals almost drowned Iowa with waste.

The animal farms of Iowa are part of a larger problem: that of massive toxic spraying of crops. The two together increase the incidence of cancer and other diseases. For example, the effect of massive pesticide pollution is certainly neurological disease and cancer because the most used pesticide chemicals in Iowa and the rest of the country are neurotoxins and carcinogens. We have cited the evidence in chapters 2–4, 6, and the conclusion of this book. I also cited the government's evidence in my book, *Poison Spring* (cited in the footnotes). Yet the Iowa model of raising animals for food and industrial farming has been the model of agriculture and animal farms everywhere. The consequences are dramatic. About half of the children living near animal farms are suffering from asthma.[35] Meanwhile, the owners of Concentrated Animal Feeding Operations (CAFOs) shoot manure and other animal farm waste into the air and land.[36] Unfortunately, neither Democratic nor Republican administrations have banned these practices, which complement the irresponsible spraying of pesticides over vast crop acreage, and directly and indirectly

---

[34] Chris Jones, Iowa's real population, Blog, March 14, 2019. https://cjones.iihr.uiowa.edu/blog/2019/03/iowas-real-population.

[35] James A. Merchant *et al.*, Asthma and farm exposure in a cohort of rural Iowa children, *Environmental Health Perspectives*, March 2005, **113** (3), pp. 350–356.

[36] Dirk Johnson, Pork producer settles suit as pollution rules tighten, *New York Times*, August 16, 1999.

contribute to the poisoning of the country and its food and drinking water.[37,38]

According to an undated article (but probably written in the 2010s) by the People for the Ethical Treatment of Animals (PETA), an animal welfare organization, factory farm animals are flooding the country with huge amounts of toxic and pathogenic wastes. "Animals on factory farms," said PETA, "generate many times the amount of excrement produced by the entire U.S. population, and this waste pollutes the air we breathe and the water we drink. Every second, our nation's factory farms create roughly 89,000 pounds of waste, which contains highly concentrated chemical and bacterial toxins – all without the benefit of waste-treatment systems."[39]

In 2010, the CDC confirmed PETA's fears. It issued a study that concluded that CAFOs "can cause a myriad of environmental and public health problems."[40] The study reported that the air surrounding these animal factories is full of hazardous chemicals and pathogens. The chemicals include ammonia, hydrogen sulfide and methane. Particulate matter adds to the toxic soup of air pollution. "[A]ll of which," the CDC said, "have varying human health risks." For example, breathing CAFOs' air threatens the lungs, skin and eyes. These dangers are heightened with disease-causing bacteria and viruses we call pathogens (vectors of disease). These pathogens emerge from the manure in the CAFOs. They cause a variety of diseases, some of which can be fatal. Common CAFO diseases and disease symptoms include:

- *Anthrax*: Skin sores, headache, fever, chills, nausea, and vomiting.
- *Colibacillosis, Coliform mastitis*: Diarrhea and abdominal gas.
- *Leptospirosis*: Abdominal pain, muscle pain, vomiting, and fever.
- *Listeriosis*: Fever, fatigue, nausea, vomiting, and diarrhea.

---

[37] E. G. Valliianatos with Mckay Jenkins, *Poison Spring* (London: Bloomsbury Publishing).

[38] Robert van den Bosch, *The Pesticide Conspiracy* (California: University of California Press).

[39] People for the Ethical Treatment of Animals, Other health risks of the meat industry. https://www.peta.org/issues/animals-used-for-food/health-risks-meat-industry/.

[40] Carrie Hribar, Understanding concentrated animal feeding operations and their impact on communities, National Association of Local Boards of Health, US Centers for Disease Control and Prevention, 2010.

- *Salmonellosis*: Abdominal pain, diarrhea, nausea, chills, fever, and headache.
- *Tetanus:* Violent muscle spasms, lockjaw, and breathing difficulties.
- *Histoplasmosis:* Fever, chills, muscle ache, cough rash, joint pain, and stiffness.
- *Ringworm*: Itching, and rash.
- *Giardiasis*: Diarrhea, abdominal pain, abdominal gas, nausea, vomiting, and fever.
- *Cryptosporidiosis*: Diarrhea, dehydration, weakness, and abdominal cramping.

Air and water usually become the invisible networks of pathogens infecting people and animals, according to the CDC study. This contagion increases dramatically in the extremely crowded conditions of CAFOs. **"Healthy or asymptomatic animals,"** said the CDC, **"may carry microbial agents that can infect humans, who can then spread that infection throughout a community, before the infection is discovered among animals"** (emphasis mine).

It is not just my own belief when I said that animal farms are factories that spark diseases with the potential to become pandemics. The environmental organization Natural Resources Defense Council (NRDC) has been warning of the dangers of animal farms for decades. In 2001, it reported that the waste of animal farms is "threatening our health, the water we drink and swim in, and the future of our nation's rivers, lakes, and streams."[41] In 2019, it stressed on the harm of drinking water contaminated by nitrates derived from manure. In the report, it was mentioned that the effect of contaminated water can be especially deleterious to very young children and fetuses. Water contaminated by nitrates from the waste from animal farms have been known for causing "blue-baby syndrome" (also known as cyanosis) – a dangerous and sometimes fatal medical condition in which the babies have insufficient oxygen in their blood. [42]

The CDC report cited above, also mentioned that living near CAFOs is dangerous because of the unbearable stench and outright harm from

---

[41] Robin Marks, Cesspools of shame: How factory farm lagoons and spray fields threaten environmental and public health, *NRDC*, July 2001.

[42] Valerie Baron, Big Ag is hiding in plain sight and it's making us sick, *NRDC*, September 23, 2019.

clouds of chemicals, pathogens, and toxic dust. In a NDRC interview, a farmer, Neil Julian Savage, who lived next to a hog CAFO in Bladen County, North Carolina, said he lived with his wife on his farm since 1952. In 1991, a hog factory belonging to Brown's-Smithfield went into business right next to his farm. This animal factory had a lagoon and ten barns. The managers of the animal factory would spray countless amount of manure and animal waste into the heavens and on their land, adjacent to his land. "Often times, even when they are not spraying hog waste," Savage said to the NRDC, "the smell from the barns and lagoon gets so bad I can smell it in my house with all the doors and windows shut. During these times, it is impossible to stay outside for even short periods. When Brown's is spraying hog waste on the fields, especially when it is near my property, living here is almost impossible. The overall situation has been so bad that I have not been able to farm my land for some time... My wife and I have been made sick by the rancid odors that are forced upon us. If Brown's is spraying near my house I cannot stand to be outside for more than a few minutes. It makes me so sick that I have fallen to the ground and had to crawl back to the house on several occasions. The same thing has happened to my wife. Sometimes it is so bad that my wife and I feel like giving up."[43]

The pain, anguish and desperation of this farmer is just the suffering of one out of millions of Americans unfortunate enough to live in the toxic neighborhood of CAFOs. Reports of CDC and NRDC cited above, documented the effects of these gigantic animal farms. The consequences are telling and terrifying. It seems to me that the potential but hidden sources for future pandemics are all over the United States, in thousands upon thousands of CAFOs. Speaking from my 27-year experience of working for the federal government, I feel confident to say that the US government has been turning a blind eye, allowing these festering disease factories to go on. The result is systemic harm in my view. Despite the grave risks to both animals and people, the owners of these large animal feeding operations refuse to revamp and upgrade their operations to manage the colossal amount of wastes from their factories. I already cited the 2020 decision of the Trump administration to include these meat producing factories as necessary businesses that had to remain open, pandemic or not. One would expect them to, at the very least, treat the animal wastes like cities

---

[43] Robin Marks, (2001), *Op. cit.*

treat human wastes. But no, they pour all those countless tons of filth and disease into lagoons and rivers causing unfathomable ecocide.[44]

The stench from those wastes is affecting primarily powerless and, usually, minority communities neighboring animal farms. This is especially blatant in east North Carolina where black Americans and other minorities live not far from millions of pigs confined for feeding and slaughter in giant hog factories.[45] I visited a black community in east North Carolina in the 1990s. Then, I was participating in an academic conference highlighting the politics and health effects of CAFOs in the neighborhood of black Americans. I heard lots of angry rhetoric then, but unfortunately, I did not hear of realistic plans to end that oppression and hazard.

CAFOs are equally dangerous to wildlife. Their waste lagoons become death lakes for flying and migrating birds. During storms, waste lagoons overflow into creeks, rivers and ground water aquifers – harming both wildlife and humans.[46,47]

In April 2009, the editor of the *Scientific American* issued his warning[48] that Americans are losing control of their "great agricultural machine." He complained that farms are ill and make us ill as well. He said orange groves in California and Florida are "falling to fast-moving blights with no known cure" while "our agricultural practices are leading directly to the spread of human disease." He cited the case where, in 2005, an antibiotic-resistant strain of Staphylococcus bacteria killed about 19,000 Americans. According to the report in the *Scientific American*, the FDA found that 49% of factory pigs and 45% of the pig workers harbored the disease's bacteria. Yet, the FDA refused to test all farm animals

---

[44]Costner and Thornton, *We All Live Downstream* (Washington: Greenpeace USA), pp. 1–4.

[45]Waterkeeper Alliance, Exposing fields of filth: Factory farms disproportionately threaten Black, Latino, and native American North Carolinians, July 30, 2020. https://waterkeeper.org/news/update-exposing-fields-of-filth/.

[46]Valerie Baron, Big Ag is hiding in plain sight and it's making us sick, *NRDC*, September 23, 2019.

[47]Amanda Day, The effect of factory farming on the environment, PETA, November 14, 2008. https://www.peta.org/news/winning-heliumcom-essay-effect-factory-farming-environment/.

[48]Editorial, Our sick farms, our infected food, *Scientific American*, April 2009. https://www.scientificamerican.com/article/sick-farms-infected-food/.

for disease. In chapter 5, I have discussed and cited evidence that factory farm workers routinely feed antibiotics to millions of stressed animals with the result of threatening the health of both animals and people, especially the millions who eat those animals. The editor of the *Scientific American* also mentioned that in 2008, a commission urged the FDA to phase out this abuse of antibiotics and the FDA promised to do so only to abandon the idea almost immediately. The infuriated editor said: "This is just one example of the food production system that protects a narrow set of interests over the nation's public health."

The Pandemic of 2020

This truth became clear in 2020 when America and the world were forced to their knees by the coronavirus pandemic. Trump and the Republicans rushed huge amounts of money to corporations and the richest Americans,[49] while largely ignoring unemployed workers living – according to Senator Bernie Sanders[50] – from paycheck to paycheck. In addition, the Trump administration was alleged to have mismanaged the country's virus health crisis, spreading misinformation and preventing a normal and responsible handling of the pandemic.

The Chimeras of Biological Warfare

The pandemic, however, influenced both domestic and foreign policy in America. In Chapter 4, I discussed why Americans blamed China for the coronavirus pandemic, as well as how and why the University of North Carolina scientists taught Chinese scientists to manipulate viruses, while leaving no traces of their genetic engineering methods. Such weaponized science brought out of darkness microscopic pathogenic monsters known as Chimeras. This danger brings us back to the controversy of which country started the pandemic, probably by accident. China or the US?

---

[49]David J. Lynch and Jeff Stein, Trump's coronavirus plan includes industry bailouts that Republicans once opposed, *Washington Post*, March 18, 2020.
[50]Donna Borak, Sanders says half of Americans are living paycheck to paycheck, *CNN*, July 30, 2019.

Milton Leitenberg,[51] analyst at the Center for International and Security Studies, University of Maryland, blamed China for the pandemic of 2020. He said that, in 2015, the Wuhan Institute of Virology in China started experimenting on how to make a natural virus more pathogenic and easier to transmit. Experts call this manipulation of viruses "gain of function." Lab manipulated viruses, said Leitenberg, could become part of lab aerosols, moving from experimental animal to experimental animal, as well as to lab scientists and technicians. "In other words," Leitenberg insists, "gain of function techniques were used to turn bat coronaviruses into human pathogens capable of causing a global pandemic." Leitenberg reviewed the published literature on the coronavirus and China. He said that there was a record of poor safety and viruses escaping from Chinese labs, at least since 2004. This includes the Wuhan Institute of Virology. Leitenberg alleged that China suppressed information, especially about the coronavirus. In addition, he insisted that China initiated a "disinformation campaign" regarding the "origins" of the coronavirus, "targeting US biological laboratories." Moreover, Leitenberg stated that the Wuhan Institute of Virology indulged in "gain of function research," which included infesting a lab animal with a bat virus. However, in my opinion, these charges against China are almost impossible to prove, and while Leitenberg admitted as much, the practice of virus manipulation is indeed a dangerous game to play.

US virology labs are probably no better than Chinese labs. However, reporters from the *Los Angeles Times* suggested that China has been muzzling researchers, which could be inferred as proof of potential Chinese origins of the pandemic. Reporters also alleged that Chinese propaganda has been promoting "fringe theories" that the 2020 plague originated among people eating frozen seafood in other countries, not China.[52] I found these arguments unconvincing.

Two other American scientists, Jonathan Latham, and Allison Wilson made a case[53] against China that is slightly more plausible than that of

---

[51] Milton Leitenberg, Did the SARS-CoV-2 virus arise from a bat coronavirus research program in a Chinese laboratory? Very possibly, *Bulletin of the Atomic Scientists*, June 4, 2020.

[52] Dake Kang, Maria Cheng and Sam McNeil, China muzzles research into virus origins, *Los Angeles Times*, January 2, 2021.

[53] Jonathan Latham and Allison Wilson, A proposed origin for SARS-COV-2 and the COVID-19 Pandemic, *Independent Science News*, July 15, 2020.

Leitenberg and the reporters of the *Los Angeles Times*. Latham and Wilson said in July 2020 that, in 2012–2013, Zheng-li Shi, a Chinese virologist, mined the "presumptive ancestors" of the coronavirus from a cave in the Yunnan province of China. Shi has a laboratory at the Wuhan Institute of Virology. In the spring of 2012, just before Shi brought to her lab the viruses that became the potential pandemic coronavirus, six Chinese miners came down with the fatal contagion. Latham and Wilson theorized that the miners' disease of 2012 became the global pandemic of 2020. They say that Shi was funded to explore the pandemic pathogenicity of the coronavirus she extracted from cave bats. They argued that the research Shi carried out in her lab, and a potential lab accident, were the causes for the pandemic.

In April 2020, Li-Meng Yan, a Chinese medical virologist and whistleblower, fled China for the United States.[54] She accused the Chinese government of suppressing the origins of the pandemic in Wuhan.[55] She alleged that far from being "natural," the COVID-19 virus was either a modified version of the bat virus or it was entirely a lab construct – a synthetic virus capable of causing a pandemic. She argued that the 2020 pandemic virus "shows biological characteristics that are inconsistent with a naturally occurring, zoonotic virus." In my view, even this plausible theory is only a sophisticated speculation.

Scientists from the University of North Carolina employed genetic engineering for the creation of a pandemic virus that they injected into a mouse, which then bit a researcher. These deadly events took place sometime between April 1st and May 6th of 2020. They further illustrate the dangers[56] of the weaponization of science and the grave risks of lab leaks. The scientists of this lab at the University of North Caroline were partners with the virologists at the Wuhan Institute of Virology in China.

Richard Ebright, virology expert from Rutgers University, warns that too many researchers, both in America and the world, are fiddling with pandemic viruses. He challenged those who opposed the lab leak theory

---

[54]Li-Meng Yan, Wikipedia. https://en.wikipedia.org/wiki/Li-Meng_Yan.

[55]Li-Meng Yan *et al.*, Unusual Features of the SARS-COV-2 Genome, *Zenodo*, September 14, 2020.

[56]Alison Young and Jessica Blake, Near Misses at UNC Chapel Hill's high-security lab illustrate risk of accidents with coronaviruses, *ProPublica*, August 17, 2020.

of the origins of the pandemic.[57] Another expert, Edward Hammond, agrees with Ebright. He works at Prickly Research in Austin, Texas. He is also a former Director of the Sunshine Project, an NGO that tracked the post 9/11 expansion of the US Biodefense program. Hammond[58] believed that the "swarms" of mostly inexperienced academic researchers jumping into coronavirus investigation might bring more trouble than light. He said that the first SARS virus came out of laboratory-acquired infections. He feared that "there is a very real risk that modified forms of SARS-CoV-2 could infect researchers… with unpredictable and potentially quite dangerous results. The biggest risk is the creation and accidental release of a novel form of SARS-CoV-2 – a variant whose altered characteristics might undermine global efforts to stop the pandemic by evading the approaches being taken to find COVID-19 vaccines and treatments."

My guess is that the Cold War provided the historical context for the 2020–2022 pandemic. The hostility between superpowers America and Soviet Union (Russia) brought the factory in the forefront of human development – much more than it was during the industrial revolution of the nineteenth century. The factory became industrialized farming – an approach which basically meant putting all our food eggs in the same basket. This demanded faith in the science of pesticides, which did not exist then or now.

As I have argued in this book with cited evidence, decades of pesticides and synthetic biology have been having their effects: poisons in the food Americans and other people eat have shattered their immune defenses, and in consequence, those with the most compromised health were killed in the 2020 pandemic. I think, unfortunately, that the inconvenient truth is lost to or intentionally overlooked by officials, politicians and medical doctors captured by the gold-laden lobbyists of the pharmaceutical, petrochemical, agribusiness, and animal feeding operations industries. These companies have become sophisticated in hiding the risk of their products. That is the main reason of lobbyists. They speak to usually not well informed politicians and convince them that they represent companies of great reputation and integrity, including scientific integrity. They fail to remind the politicians that for decades, chemical companies

---

[57]Michael Hiltzik, Two Rutgers professors are accused of poisoning the debate over COVID's origins, *Los Angeles Times*, March 20, 2024.
[58]Jonathan Lathan, Engineered COVID-19-infected mouse bites researcher amid "explosion" of risky coronavirus research, *Independent Science News*, August 13, 2020.

used fraud in support of the pesticides they brought to the US EPA for approval. I documented that fraud in my book, *Poison Spring*. In fact, sometimes companies and their lobbyists manage to get the government to fund the disempowerment of those revealing the risks of their products.[59]

Second, I am convinced that all these conflicting and dangerous developments run parallel to genetic engineering and biological warfare. Biological warfare in the name of biosafety and biodefense reminds us of similar claims of advocates of nuclear warfare. In my opinion, the pandemic of 2020 was no different from a nuclear weapon exploding by accident. Who is going to admit the gravity of such an accident? Or dare to suggest, as I have done in this book, that a version of biological warfare has been taking place, invisibly to a large extent so far, at thousands of animal farms for decades? I maintain that the viruses at animal farms are potentially coronaviruses.

The Animal Factor

While we may set aside the potential errors of biological warfare, the genetic engineering of viruses, and lab accidents, in my opinion, we cannot ignore the most likely avenue of pandemics: their agricultural origin. In Chapters 2–4 and the beginning of this chapter, I suggested that pesticides are at the center of the Covid-19 and future pandemics. I believe that there's little doubt that CAFOs – made possible by pesticides and one-crop farming – have, in many cases, been fertile grounds for animal suffering and human pestilence. For example, in a video posted on the blog of Mercy for Animals, pregnant sows are locked into iron crates small enough that prevent the sows from getting up or turning around. Their waste goes down holes under their crates, though the animals are engulfed by the ammonia of those wastes. Moreover, this violence against the sows brings about diseases harming their legs and feet. Sows also desperately keep biting the iron bars of their cages.[60] Once the piglets are born, the sows are forced into other "farrowing" crates where for a few weeks they

---

[59] C. Gillam, M. Gibbs and E. DeBre, Defend or be damned – How a US company uses government funds to suppress pesticide opposition around the world, *The New Lede*, September 27, 2024.

[60] Hannah Bugga, Pigs Driven Mad by Extreme Confinement, Mercy for Animals, August 5, 2023; a brief video by Mercy for Animals shows the violence against sows. https://mercyforanimals.org/blog/new-footage-pigs-driven-mad/.

feed their baby piglets. Then the violence of extreme confinement starts all over again. Artificial insemination, tiny crates, immobility, and the breathing in of ammonia. "Constricting animals in cages," said an editorial of the *Los Angeles Times*,[61] "is one of the cruelest practices in factory farming, which is also perpetrated on egg-laying hens and veal cattle."

The NRDC[62] highlighted some startling facts about the prevailing daily violence at CAFOs. America raises and largely eats more than 9 billion animals per year. These animals live their short lives in "massive industrial facilities" in "horrific conditions." In addition, they are forced into gas chambers where they are asphyxiated by carbon dioxide on their way to have their throats slit.[63] The animal slaughter industry has been saying the gas chambers are humane. Raven Deerbrook, a courageous young woman risked her life by secretly entering the gas chambers of Smithfield Foods in Vernon, California. She placed four tiny spy cameras that taped the hogs' terror of asphyxiation by carbon dioxide. Raven shared the videos with *Wired Magazine*.[64] I watched one of those videos and I can say throwing hogs or other animals into gas chambers is the equivalent in pain and horror to having fish out of water. I observed animals screaming, squilling, moving their bodies violently, keeping their mouths open for air, and finally dying the most horrible death imaginable.

The violence of the animals' gas chambers extends outside the chambers. I have cited studies by the CDC, the NRDC, the Sierra Club, PETA and Mercy for Animals documenting the dangers and harms of CAFOs. Animal farms pollute the air, water, and land on a massive scale. The pollution of each CAFO harms humans and nature alike. Even though there are thousands of animal factories all over America, NRDC noted that they remain unregulated. Not even the EPA knows the exact number

---

[61] Editorial, Walmart hasn't stopped pork suppliers from confining pregnant sows in cruel cages, *Los Angeles Times*, May 31, 2024.

[62] Valerie Baron, Big Ag is hiding in plain sight and it's making us sick, *NRDC*, September 23, 2019; lmart.

[63] Nicholas Kristof, Spy cams show what the pork industry tries to hide, *New York Times*, February 4, 2023.

[64] Andy Greenberg, Spy cams reveal the grim reality of slaughterhouse gas chambers, *Wired*, January 18, 2023. https://www.wired.com/story/dex-pig-slaughterhouse-gas-chambers-videos/.

and location of CAFOs.[65] My guess is that fear supports secrecy. Perhaps senior government officials do not wish to offend the owners of these big animal farms. The CDC[66] warned us: put thousands, much less millions, of animals very close to each other and you guarantee disease among those animals and disease among the humans feeding and slaughtering and processing the meat of those animals. Yet, as NRDC[67] also warned: many urban Americans know practically nothing about these rural factories; out in the open for all to see, but invisible, nevertheless. In my opinion, these factory owners are aware that if people in the neighborhood of the factories and those in the cities, discover the diseases behind the pigs, chicken and cattle, they will be in trouble.

Frankenstein's Chimeras

Writing in 2013, Indiana University professor J. E. Hollenbeck[68] said that there is danger in expanding CAFOs to include rice–duck farming and pig–duck–fish aquaculture. The danger is in multiplying mischief in pandemic diseases. This is especially true for the poultry CAFOs that are 10-million-bird-mega-farms, which make possible genetic drift among confined animals and possibly human workers. When you have numerous pig and chicken factories so close to each other, "the acceleration of the mixing and assortment of influenza viruses is unfathomable," Hollenbeck said. He worried that human, swine, and poultry viruses would spread. His solution? Monitor the livestock and the workers feeding animals. And add the live animal markets of Southeast Asia to monitoring to prevent a repetition of a pandemic, he said.

Monitoring animal markets in Asia and all over the world, including animals and people in CAFOs in America is important. However, I believe that a neutral party is needed to do the monitoring as public health is crucial for the world.

I have been arguing that the best defense against another pandemic is to stop the destruction of the natural world; dismantle, worldwide, the

---

[65] Valerie Baron (2019), *Op. cit.*

[66] Carrie Hribar (2010), *Op. cit.*

[67] Valerie Baron (2019), *Op. cit.*

[68] J. E. Hollenbeck, Concentrated animal feeding operations (CAFOs) as potential incubators for influenza, *Trakia Journal of Sciences*, (2), 2013, pp. 205–209.

biological warfare laboratories, and end the genetic engineering and agricultural causes for global pestilence: pesticides, industrial agriculture, and animal farms. I firmly believe that badly treated animals are at the core of these tragedies and that the political, chemical, and biological technologies behind animal farms have given rise to a Frankenstein's Chimera. I theorize that this Chimera is taking the shape of ever-changing genetic mixtures of different species that do not belong in the natural world or among people. These viruses can be almost immortal. They will threaten the human race with disease and death. In the event that such a biological man-made Chimera really exists, it would be infinitely worse than the fictional mechanical man-monster killing machine in Mary Shelley's 1818 novel *Frankenstein*. Virologists appear to trivialize their creations with the innocent code words like "genetic swap," "gene stuffing," or "recombination," but, in my professional view, any virus-Chimera birthed from lab or animal farm viruses can become a ticking bomb, not much different than the streaming fire of the original Greek mythical monster. The latter was described by the late eighth–early seventh century BCE epic poet-philosopher and bard Hesiod as a mighty, very large, dreadful, fleet-footed, female monster that breaths fire ceaselessly. She is described to have three heads – one of a shining lion, another of a goat and still another (the third) of a savage snake. The front of her body was likened to that of a lion, her tail, a dragon, and her upper-middle as a fire-breathing goat.[69] I think the fire of the mythical Greek Chimera in the modern, biological man-made Chimera has taken the form of disease and death.

Disarmament

Biological disarmament – i.e., ceasing the creation of more Chimeras – should be no less important than nuclear and chemical weapons disarmament. I hope we can eliminate pesticides, industrial farms, and CAFOs. They would be the logical and necessary reform steps to ease the risks of pandemics from thousands of giant crop and animal farms in America – and additional thousands elsewhere on the planet. America and the world without pesticides, industrial farms, and CAFOs would be safer and healthier in my opinion.

---

[69] Hesiod, *Theogony*, 319–324.

The last chapter – conclusion – argues that we should not slaughter animals for meat – an act I believe to be immoral, extremely violent, and unnecessary. Moreover, as shown earlier, CAFOs emit great amounts of greenhouse gases; banning them would be a blessing for our fight against climate change – by far the most ominous threat and danger to human survival and the survival of the natural world.

Conclusion

Photo of a painting of Pythagoras advocating vegetarianism,
by Peter Paul Rubens (1618–1620). Public Domain.

In the Village

I am adding some more biographical details to connect this book even more to my upbringing, education, and experience. As mentioned in the introduction, I grew up in a Greek village. My most interesting memory comes from our harvesting of grapes during the heat of summer – typically around late August. My sisters and cousins would fill wicker baskets with ripe bunches of white, blue and red grapes, and load them on the donkey, and my younger cousin, George, and I would take them home. We would unload the baskets and pour the grapes into *linos*, a rectangular stone and cement enclosure 1 meter high, 3 meters wide, and 6 meters long. One of the stone walls of the *linos* had a hole that allowed the liquid wine to drain into a small cement pit below. After filling the *linos* with the ripe and tasty fruits, George and I would wash our legs and enter the soft hills of grapes, which we treaded to pulp while laughing, singing, and having fun.

In America

At age eighteen, I left the village for America where I discovered more of the beauties and pleasures of Greek civilization – and much, much more. This happened slowly. Like other young Greeks and most foreign students from many countries, I saw America as a land of opportunity for those with technical knowledge and skills. This pushed my love for the Greek classics into the recesses of my mind. In 1961, when I arrived in America, I simply wanted some education that would enable me to earn a good living. I had a vague notion of a good life. However, my education in zoology, Greek history, and the history of science coupled with my work on Capitol Hill and the US Environmental Protection Agency brought me face to face with modernity – and I did not like it. I could not stand looking at skyscrapers and cringed at seeing gigantic tractors crushing the land. I had the feeling I had to turn to classical Hellenic thought. If I were to survive, I would have to have the support of my ancestors.

I read the writings attributed to Pythagoras[1,2] with great interest. Pythagoras was a sixth century BCE polymath: mathematician, astronomer, cosmologist, musician, theologian, ecologist, and advocate

---

[1]Kenneth Sylvan Guthrie, tr., and David Fideler, ed., *The Pythagorean Sourcebook and Library* (Grand Rapids, MI: Phanes Press), 1987–1988.
[2]Charles H. Kahn, *Pythagoras and the Pythagoreans: A Brief History* (Indianapolis, IN: Hackett Publishing Company), 2001.

of vegetarianism. He was a philosopher of the heavens and Earth. He believed that Number was the key to the Cosmos: a constituent of everything. He thought music and songs had a healing and educational effect, invigorating humans with inner strength and harmony. He even said he heard the music of the spherical planets moving around the Sun, which he equated to a large fire at the center of the Cosmos. He called that fire the House of Zeus. He was in love with animals and life. He was against destroying or eating any living being.

Pythagoras was certain there was a brotherhood between humans and animals. He urged the Greeks to stop eating meat and never sacrifice animals to the gods. Iamblichus, a fourth-century Greek scholar, admired Pythagoras' mathematical wisdom. "Pythagoras," he said, "triggered all the exact sciences and branches of knowledge, everything that gives the soul true vision and clears the mind blinded by other practices [like eating animals]."[3,4]

I also reread the works of Herodotos[5] and Thucydides,[6] the greatest historians of ancient Greece. Herodotos' story of the Persian Wars is memorable for its comprehensiveness, objectivity, and importance, and has been regarded as a model of storytelling ever since. Thucydides focused his story on the Greeks' civil war, known as the Peloponnesian War. Like Herodotos, Thucydides wrote for all time, digging deep into causes and consequences, thus giving us a unique study of war, and why the Greeks started killing each other. Fear was the decisive factor. The Spartans knew they were the Greek superpower – at least that's how they thought of themselves. But they did not know how to handle the perpetual rise to power of Athens. Fear of Athens and the spread of Athenian ideas and goods all over the Greek world, Thucydides concluded, brought the Spartans to the field of battle, thus triggering the Peloponnesian War. Thucydides was so distressed about the bloodshed and destruction that he did not finish his masterpiece, *The Peloponnesian War.*

I read Xenophon, an Athenian military man and historian who was a contemporary of Plato, who grew up during the Peloponnesian War. Like Plato, he was influenced by Socrates. He wrote the history of the last

---

[3] Iamblichus, *On the Pythagorean Life*, 6.31.

[4] Evaggelos Vallianatos, *The Antikythera Mechanism: The Story Behind the Genius of the Greek Computer and its Demise* (Irvine, CA: Universal Publishers, 2021), pp. 29–31.

[5] Herodotos, *The Histories*, tr. Robin Waterfield (New York: Oxford University Press, 1998).

[6] Thucydides, *The Peloponnesian War*, tr. Walter Blanco (New York: W. W. Norton, 1998).

seven years of the Peloponnesian War, a work he called *Hellenica* (*Hellenic History*).[7] He flourished in the first half of the fourth century BCE. I agree with his theory and conviction that agriculture was a school for courage, freedom, military training, and the raising of food and civilization. In *Oikonomikos (Home Economics/Management)*, Xenophon says that just like soldiers sacrifice to the gods before battle, so should wise farmers pray to the gods for the health and prosperity of their property on land and water, including horses, cattle, sheep and crops.[8] Small farmers in ancient Greek *poleis* (city-states) raised food and freedom. They volunteered in defending their *polis* and served as jurors in the government. They were intimately connected to the festivals of the polis honoring gods like Demeter, Athena, Dionysos, Aristaios and Pan who blessed wheat and other grains and supported wildlife, pollinators, grapes, wine, flocks of animals, cheesemaking and olive trees and the harvesting of wholesome food. These farmers were the *polis*.

Then the fourth century BCE philosopher Aristotle came into my life like a bolt of lightning and a breath of fresh air. In contrast to the dry and uninspiring classes I took while studying zoology at the University of Illinois, the writings of Aristotle brought me in touch with the roots of zoology. His works on animals, especially his *History of Animals* and *Parts of Animals*,[9,10] lifted me to heavens. They were insightful, riveting, enormously important, and pioneering. They explained to me the origins, complexity, and beauty of the animal kingdom, the perfection and truth of nature, as well as the meaning and importance of the science of zoology, which Aristotle invented. For example, Aristotle said nature is marvellous and divine. She created animals, which reveal nature and beauty, even the humblest of them. He urged us to study animals for knowledge and philosophical reflection. After all, he said, we all live in their midst.[11] Aristotle studied hundreds of animals. In his study of the dolphins, for example,

---

[7]Xenophon, *Hellenika*, tr. John Marincola and ed. Robert B. Strassler (New York: Anchor Books, 2009).

[8]Xenophon, *Oeconomicus*, 5.20.

[9]Richard McKeon, tr., *The Basic Works of Aristotle* (New York: The Modern Library, 2001).

[10]Aristotle, *Historia Animalium (History of Animals)*, ed. D. M. Balme (Cambridge: Cambridge University Press, 2001).

[11]Aristotle, *Parts of Animals*, 644b28–645a25.

he expressed admiration for their speed. He described them as mammals with a gentle disposition and passionate attachment to their young.[12]

At Work

I cannot say these Hellenic scientific, ecological, and philosophical insights blended nicely with my life. After a couple of years on Capitol Hill, I started working for the US Environmental Protection Agency (EPA). The year was 1979. For the first time, I began to grasp what America was all about. I was so embarrassed that my chosen new home, the United States, was not what I thought it was – the land of opportunities: I was shocked to find out that scientists at the EPA and other agencies like the US Department of Agriculture (USDA) would employ science in the "regulation" of organophosphorus pesticides, chemicals that were modified forms of chemical warfare agents used in World War 1. Sadly, science is unable to control these nerve poisons.[13,14] The only solution is not to allow such deleterious nerve gases to be used in farming or any other activity. The struggle over what to do about these pesticides – whether it is reducing the amounts of chemical that would reach water, air, and land; releasing no more gases into the environment; banning them; and/or cleaning up contaminated areas – continues. It appears to me, that the struggle is like a metaphor of the anger and passion unleashed by the agribusinessmen who heavily influence rural America and the government. It was also obvious to me that the "relaxing" of government regulations under the Trump administration reflected the desires of the regulated industry. Trump stopped the "surprise inspections of chemical and power plants." Moreover, he approved the insecticide, sulfoxaflor, that is harmful to honeybees. And in late 2020, the EPA allowed the continued use of a neurotoxin called chlorpyrifos.[15] In the report, it was alleged that the EPA ignored several epidemiological studies, especially the one from Columbia University, that had found a correlation linking

---

[12] Aristotle, *History of Animals*, 9.623b5–627b22; 631a9–631b3.

[13] Sudisha Mukherjee and Rinkoo Devi Gupta, Organophosphorus nerve agents, *Journal of Toxicology*, September 22, 2020.

[14] Evaggelos Vallianatos, Expendable endangered species, *Counterpunch*, October 4, 2023.

[15] Lisa Friedman, E.P.A. won't ban chlorpyrifos, pesticide tied to children's health problems, *New York Times*, July 18, 2019.

prenatal chlorpyrifos exposure with brain developmental disorders in children. But, it would seem that Trump and the EPA rejected science documenting that chlorpyrifos is stunting "brain development in children."[16]

Trump was not alone in favoring large industrialized farmers and the agrochemical industry. The Democratic presidents before him, Barack Obama (2009–2017), Bill Clinton (1993–2001), and Jimmy Carter (1977–1981), were not much better. For example, in 1980, I invited Professor of Entomology David Pimentel of Cornell University to speak at EPA. He gave a fantastic lecture and warned the scientists and senior officials of the dangers of the chemicals EPA approved. Also, genetic engineering for making seeds immune to herbicides were developed during the democratic administration of Bill Clinton. In addition, I have documented the EPA's regulation of pesticides from its inception till about 2010.[17] The record, unfortunately, is not one that shines in favor of public and environmental health. The 2020 pandemic added to my fury. As I have explained in this book, the pandemic added more opportunities for mischief in all sectors of politics, science and society. The fear of the disease and the fear of officials to search for the source of the viruses added to inaction and increased death. Similar politics and corruption of decades ago informed and angered me. I was confused, and more than a little concerned about this gigantic country I had chosen as my second home.

Decoding Scientific Research

Frustrated that I was unable to influence or change policy at the EPA, I turned to research and writing. Scientists publish important work but perhaps, in order to protect themselves, or because they do not have sufficient convincing data, they are often not clear in their findings and conclusions. They also tend to publish their research in obscure journals read by few people, most likely other scientists. I tracked down dozens of those research papers, and merged their key findings with the highlights

---

[16] Lisa Friedman, E. P. A. rejects its own findings that a pesticide harms children's brains, *New York Times*, September 23, 2020.

[17] E. G. Vallianatos with Mckay Jenkins, *Poison Spring* (London: Bloomsbury Publishing, 2014), pp. 1–236.

of the stories I had heard from some of my EPA colleagues, who also gave me their memos and briefings. In addition, I met a few outstanding scientists who answered some of my questions about pesticides, agriculture, animal farms, water, endangered species, biodiversity, and politics. They worked for universities or the federal departments of the Interior and Agriculture. Once I retired from the government in 2004, I explained what the EPA did with its regulation of pesticides in my 2014 book, *Poison Spring: The Secret History of Pollution and the EPA.*

Origins of the 2020–2022 Pandemic

The proliferation of the coronavirus in 2020 sharpened the political, scientific and agricultural crisis. A 2023 Report sheds some light. The Center for Environmental and Animal Protection at New York University and the Brooks McCormick Animal Law and Policy Program at the Law School of Harvard University published that Report[18] on animal diseases (zoonotic or zoonoses). In the case of emerging diseases, the Report said, nations point fingers at one another while ignoring the situation at home. It can't happen here, they argued. "Perhaps," the authors of the Report said, "nowhere is this attitude more palpable than in the United States." For example, during the second half of the twentieth century, the United States gave rise to more infectious diseases than any other country on the planet. The Report also pointed out that the 1918 Influenza pandemic found its epicenter in America's Midwest. The 1918 pandemic infected 500 million people, i.e., a third of the world population. It killed 12 times the number of people who died from Covid-19. And the 1918 Influenza killed more Americans than those who died from both World Wars and the Vietnam War. The Report also cited that the US spent 1650 times more on military defense than on pandemic preparations. In 2009, Influenza H1N1, swine flu, infected more than 100 million Americans, sending to hospitals more than 900,000. Finally, the New York University-Harvard University Report made the connection of risk and the zoononic diseases from animal farms. The risk is "staggering," in America,[19] precisely the

---

[18] Ann Linder *et al.*, *Animal Markets and Zoonotic Disease in the United States*, New York University and Harvard University, 2023.

[19] Emily Anthes, The risk is staggering, report says of disease from U.S. animal industries, *New York Times*, July 6, 2023.

thesis of this book. The Report, furthermore, said America "produces" more than 10 billion animals for food per year. This is a billion more animals than the 9 billion animals Americans eat every year, a 2019 estimate of the Natural Resources Defense Council.[20]

My story about the pandemic is a continuation of the narrative in *Poison Spring*. I explored the theories of the origin of pandemics – next to the realities of animal farms, or CAFOs, all over America. CAFOs stands for Confined Animal Feeding Operations.

As I have repeatedly argued in this book, the vast number of reports and articles published followed the official line of interpretation. COVID-19 or SARS2, they say, has two potential sources. The natural world is the first and has the most followers among scientists. I have said that this connection of the pandemic with the natural world brings out the worst in human nature. In various chapters of this book, I shared how people exploit the forests for wood, cut down trees to clear land for plantations (of monocrops like corn and soybeans), hunt wildlife for meat and, in general, push animals to extinction or make life miserable for surviving wild animals (like bats). Indeed, the destruction of ecosystems is so massive that some scientists suggest that nature may not be able to heal from the anthropogenic wounds humans inflict on her.[21] The forest exploiters – corporations, governments, petroleum companies, and desperate peasants – I would argue, are largely ignorant of how nature works. As for the large exploiters, like corporations and governments, the excuse for destruction is profit and greed. Climate change makes no difference to them. In 2023, for example, climate change exacerbated the immense forest fires in Canada. In 2023, forest destruction worldwide was responsible for the emission of 2.4 gigatons of carbon dioxide, which were about half of the US annual carbon emissions. In 2023, the world lost about 9.1 million acres of tropical forest, an area the size of Switzerland.[22]

---

[20] Jon Devine and Valerie Baron, CAFOs: What we don't know is hurting us, *Natural Resources Defense Council*, September 23, 2019.

[21] Rosemary A. Mason, The sixth mass extinction and chemicals in the environment: Our environmental deficit is now beyond nature's ability to regenerate, *The Journal of Biological Physics and Chemistry*, 2015, **15** (3), pp. 160–176.

[22] Manuela Andreoni, Global forest loss remains high, despite recent progress, *New York Times*, April 4, 2024.

Greed is central in deforestation. In late 2022, Congo put up for auction huge areas of its peatland-rich forest. But oil drilling in that huge forest promises vast destruction of wildlife and the poisoning of fish. Disturbing the peatlands with oil excavation would release about "5.8 billion tons of carbon."[23] Human greed is not limited to petroleum extraction in pristine areas of forests like those of Congo or Brazil. Humans are reaching almost to the neighborhood of bat caves and bats move to the neighborhood of farms. This is because deforestation and climate change empower each other, creating destitution among human residents of the forest and diminish food sources for bats. For example, black flying foxes (bats) seek nectar from the flowers of Eucalyptus, in Australia. However, if weather conditions diminish the flowers of the trees, the bats seek food near farms. That process of hunger and the stress that follows deprivation increase the shedding of viruses by bats. Deforestation and climate chaos seem to precipitate viral spillovers.[24]

Back to the theories of what gave birth to the pandemic. The advocates of the nature theory say the coronavirus is a zoonosis, i.e., a disease that jumped from animals to humans. The most likely animals that transferred the Covid-19 viruses to humans were the horseshoe bats. These wild animals, bats, says Ferris Jabr,[25] contributing writer for the *New York Times Magazine*, "were planets unto themselves, teeming with invisible ecosystems of fungi, bacteria and viruses. Many of the viruses multiplying within the bats had circulated among their hosts for thousands of years."

The second hypothesis blames the Institute of Virology in Wuhan, China for an intentional or accidental leak of the pandemic virus. Certainly, the biolabs in the United States, Europe and the rest of the world could do the same thing, whether by accident or intentionally.

Jeffrey D. Sachs,[26] professor of economics at Columbia University and advisor to world leaders, also suggested that the destruction of nature,

---

[23] Simon Lewis, The worst place in the world to drill for oil is up for auction, *New York Times*, July 15, 2022.

[24] Emily Anthes, Deforestation brings bat-borne virus home to roost, *New York Times*, November 16, 2022.

[25] Ferris Jabr, How humanity unleashes a flood of new diseases, *New York Times Magazine*, June 25, 2020.

[26] Jeffrey D. Sachs, Testimony, Committee on Oversight and Accountability, US House of Representatives, March 6, 2023.

and dangerous biosafety labs, may explain the origins of the pandemic. On March 6, 2023, he testified in Congress. "After more than three years of the Covid-19 pandemic," his testimony reads, "and more than 18 million deaths worldwide... we still do not know the origin of SARS-CoV-2, the virus that causes Covid-19. We do know, however, four things. First, the virus may have emerged from dangerous laboratory research. Second, this dangerous research was partly funded by the US Government, and notably NIH [National Institutes of Health], so the dangerous research involved US–China collaboration. Third, the NIH leadership and a group of scientists associated with NIH hid the possibility of a laboratory origin from the Congress and the public. Fourth, the question of the origin of SARS-CoV-2 is not a partisan issue... During 2020–2022, I chaired the Lancet Covid-19 Commission. The Lancet Commission followed the evidence on origins closely and concluded that both a laboratory origin and a natural origin are possible... The issue of the origin is very important. If indeed the virus emerged from a laboratory, it reveals **the remarkable danger of ongoing gain-of-function research** largely unknown to the Congress and public. There is **little if any oversight and accountability of such research, despite its dangers**. In fact, very recently, researchers at Boston University [(BU)] genetically reengineered the Omicron Variant of SARS-CoV-2 in a manner that made it more pathogenic. Yet BU did not have any federal risk-benefit review or oversight for that dangerous work. US-government-funded gain-of-function research of concern is apparently underway in many locations – and is occurring without the federal risk-benefit review and oversight that, on paper, are mandated by US policies but that, in practice, have been ignored by US research funding agencies. Since the US has a vast and largely secretive biodefense research program, additional US-government-funded gain-of-function research of concern may be underway to an extent that is unknown by the public and the Congress. This is not to mention the extent of such research in other parts of the world" (emphasis mine).

Sachs' suspicions are legitimate. NIH funded the Institute of Virology in Wuhan for gain-of-function research on bats, making the bat viruses potentially pandemic viruses. However, like other critics, his concerns about labs potentially leaking the deadly viruses are based on speculations. None of the nature spillover virologists, critics of potential lab leaks nor government agencies are pointing fingers at animal farms as one of the probable sources of the coronavirus pandemic. I am.

The Plight of Animals

Factory farms are sometimes described as meat processing operations. In my opinion, they are more than that. They are factories of disease and animal slaughter. In the previous two chapters, I have detailed how these factories put domesticated animals in the maws of machines, which feed, electrocute, asphyxiate and slaughter them, before selling their meat to meat-eating consumers.

Indigenous people like the Native Americans of the Plains ate buffalos for sustenance and culture.[27] I find no problem with that indigenous tradition. The buffalos were everything to the Native Americans: religion, culture, courage, and food. The buffaloes were them. They sacrificed a few of those sacred animals for food, shelter, survival and wellbeing. In addition to Native Americans, other people often ate animals because they had to; those living in mountainous regions with limited access to fishing or land suitable for growing fruits and vegetables and crops relied on raising sheep and goats. Ancient Greeks, for example, ate primarily wheat and barley bread, cheese, olive oil, fruits, nuts, and vegetables, and every so often, they ate fish and the meat of cattle, sheep and goats, and even sacrificed oxen, sheep, and goats to their gods.[28,29,30,31] We did the same thing in my family, eating sheep or goat meat once a month, but never eating cattle meat. In contrast, modern animal farms completely dissolve any contact people have had with animals or the natural world. The mechanization of animal death in animal farms is so inhuman that makes the slaughtered animal not a being but a thing completely unrelated to humans, not even an animal anymore. This practice alienates humans from the natural world that has been giving us birth and sustainance.

According to the non-profit organization, Animal Welfare Institute,[32] the fate of turkeys in America is horrific. Americans eat about 50 million

---

[27] *The American Buffalo: A Film by Ken Burns*, PBS, 2024. https://www.pbs.org/kenburns/the-american-buffalo/.

[28] Hesiod, *Works and Days*, 230–240, 460–490.

[29] Hesiod, *Shield*, 285–305.

[30] John Wilkins and Shaun Hill, tr. and (ed.), *Archestratus: Fragments from The Life of Luxury* (Devon, UK: Prospect Books, 2011).

[31] James Davidson, *Courtesans and Fishcakes: The Consuming Passions of Classical Athens* (New York: St. Martin's Press, 1998), pp. 3–49.

[32] Animal Welfare Institute, Grim fate for factory-farmed turkeys, *AWI Quarterly*, Winter 2020, **69** (4), p. 10.

turkeys every Thanksgiving and another 22 million at Christmas. These birds are raised in the dim darkness and crowdedness of animal farms. They do not get to breathe the fresh air of the outside world nor enjoy the light and green grass. Their sole job is to eat. To minimize the aggression of these caged turkeys, workers cut off their beaks and toes, resulting in skeletal and leg deformities that cripple and kill turkeys. On top of that cruelty, perpetual eating also results in obesity in the birds.

I have already mentioned that I have watched videos and documentaries that showed how animal factories kill their livestock. Various methods such as gas chambers and mechanical slaughter are used regularly. Mechanizing the slaughter of animals is probably the worst of human violence against animals. The suffering of the slaughtered animals is "mind-melting in its scale." Some 80 billion animals the world over are slaughtered yearly and about 41 to 160 billion farmed fish have the same fate.[33] When I watched those videos and documentaries featuring the killing of livestocks, I can only say that the experience was horrific and barbaric. It's horrific seeing a man putting an electric gun on the head of entering cattle or seeing animals entering a gas chamber and observing the sheer terror of the asphyxiation.[34] These terrifying practices undermine the philosophical and biological and agrarian connections humans have had with the natural world.[35]

Gaming the System

Large farmers and ranchers game the system with their money power, reducing our efforts to protect human health and the health and integrity of the natural world.[36] We see that in the laws defining and protecting organic food – meaning food raised without synthetic chemicals – the use of sludge as fertilizers, and the irradiation of food. Similar practices affect

---

[33] Ezra Klein, We will look back on this age of cruelty to animals in horror, *New York Times*, December 16, 2021.

[34] Glen Greenwald, Hidden video and whisleblowers reveal gruesome mass-extermination methods for Iowa pigs amid pandemic, *The Intercept*, May 29, 2020.

[35] See *The Ghosts in Our Machine*, a documentary on the treatment, plight and slaughter of animals in the laboratories and animal farms. Directed by Liz Marshall, Bullfrog films, November 8, 2013.

[36] Cara Buckley, Life [of Poultry Farmers] after factory farming: The longer they're out, the happier they are, *New York Times*, August 29, 2024.

the raising of natural crops; money power is also evident in laws designed to prevent pollution of the air we breathe and the water we drink, and laws protecting endangered species.

I have been looking at agriculture for inspiration and to understand the anthropogenic changes altering our planet and our society. Unfortunately, my views remain that industrialized agriculture and animal farms are far from inspiring. I have argued that this form of raising food, and the cruel treatment of animals, represent the worst form of farming.[37,38] I have also documented in this book that factory agriculture produces social and ecological desolation in rural America. In addition, food grown with neurotoxic and carcinogenic pesticides cause diseases and poisonings. Factory farming also releases huge greenhouse gas emissions (that escalate climate change), and increase the probability of encountering new pandemic-capable zoonotic viruses.

Chapters 2 to 7 demonstrated that the conversion of traditional small family farms into factories in the field was a very bad and irresponsible idea and policy. The new conglomerate, a Frankenstein and chemical warfare hybrid, metamorphosed into the disease Chimera that became the COVID-19 pandemic. In my opinion, the disease is connected to two sources. The first is pesticides: Paul Ehrlich, biology professor at Stanford University, said "pesticides are an ideal product: like heroin, they promise paradise and deliver addiction. And dope and pesticide peddlers both have only one cure for addiction: use more and more of the product at whatever cost in dollars and human suffering (and in the case of pesticides, in environmental degradation)."[39] Moreover, pesticides cause a multitude of diseases. They are biocides that kill and devastate wildlife.[40,41,42,43] The second source of the pandemic is animal farms: modernity's gigantic meat factories causing suffering and diseases for billions of domesticated animals. Moreover, animal farms are large sources for Earth-heating gases.

---

[37] A. V. Krebs, *The Corporate Reapers* (Washington D.C.: Essential Books, 1992).

[38] E. G. Vallianatos, *This Land is Their Land* (Maine: Common Courage Press, 2006).

[39] Paul Ehrlich, Preface, in *Bosch, The Pesticide Conspiracy*, pp. xv.

[40] Rachel Carson, *Silent Spring* (Boston: Houghton Mifflin Harcourt, 1962).

[41] Carol van Strum, *A Bitter Fog* (San Francisco: Sierra Club Books, 1983).

[42] Robert van den Bosch, *The Pesticide Conspiracy* (Berkeley, CA: University of California Press, 1989).

[43] E. G. Vallianatos with Mckay Jenkins, *Poison Spring* (London: Bloomsbury Publishing, 2014).

When the COVID-19 pandemic hit America, it brought out all my secret concerns, indeed, nightmares I had about pesticides and animal farms. I could imagine the the viruses striking people already vulnerable from eating pesticide-tainted food. So, I started writing this book as a warning to all and a note to my grandchildren. The implications of the 2020 pandemic will reverberate far into the future – very much like its predecessors. The fourteenth-century Black Death pestilence and the 1918 World War I Influenza epidemic sped up the transformation of Europe into a more centralized economy and industry. I believe that the COVID-19 pandemic and its successors are likely to exacerbate further the chasm between rich and poor, and disrupt efforts to put a brake on climate change. The hesitation to put a brake on climate change may result in the potential annihilation of human civilization, and spell the end of humanity and life on Earth.

My personal view is that China is a giant factor in the future of humanity or its destruction due to its large but still industrializing agricultural landscape. What it does is likely to impact the rest of the world. During the pandemic, China faced the same challenges as other industrial countries – particularly the dilemma to initiate lockdown or continue with work as usual. The restrictions of lockdowns, mandatory quarantines, and mass testing for Covid triggered dissent and civil unrest in large cities[44] The *New York Times* reported that Americans saw, and would have liked to see, more political implications to the COVID policies of China – perhaps Chinese citizens challenging the power of Xi Jinping and the Communist Party: "COVID-19 has shattered the unspoken social contract between the Communist Party and its people: stability and prosperity, in exchange for a high degree of social control," wrote the *Times*.[45]

Ban Pesticides and Drugs

Are we to cooperate in fighting this ecumenical climate chaos? I have mentioned in various chapters earlier, that Guterres, the Chief of the United Nations, has been warning world leaders that their childish

---

[44]Laureen Jackson, China's dramatic dissent, *New York Times*, November 30, 2022.
[45]Editorial Board, What the Chinese people are revealing about themselves, *New York Times*, December 3, 2022.

nationalisms, militarism, imperial nightmares, strategic competition, and capitalisms or socialisms ignore the climate chaos fed by humanity's burning of fossil fuels for more than a century. The existence of a global climate emergency demands that prime ministers and presidents start taking urgent actions. On February 6, 2023, Guterres told them, "End the merciless, relentless, [and] senseless war on nature."[46]

We can start by eliminating pesticides from our modern farming practices. The introduction of these chemicals to human society and the natural world more than a century ago has been a highly contested terrain. It was an act of war against insects.[47] It meant ecocide.[48,49] And it was also a serous threat against human health.[50,51] I have spoken vigorously for several decades against them because – with the exception of a few public health protection chemicals – I found no good reason for the existence of most pesticides or their use in agriculture. I document this inconvenient truth in the book I wrote after 25 years of work at the US EPA.[52] I wrote that book primarily from data I extracted from government documents, and from what I learned during my 25-year service at EPA.

I believe that Americans want to see an EPA which is a real environmental protection organization. America needs to fashion an EPA independent of presidents and their political appointees as well as corporate influence. We must make of EPA, a Federal Reserve-like organization, defending nature and public health.

---

[46] *UN News*, February 6, 2023.

[47] Will Allen, *The War on Bugs* (Vermont: Chelsea Green Publishing, 2008).

[48] Rachel Carso, *Silent Spring* (Boston: Houghton Mifflin Harcourt, 1962).

[49] Beyond Pesticides, "Threatened Biodiversity and Ecosystems," *Pesticides and You*, 2023–2024, pp. 87–117.

[50] Carol van Strum, *A Bitter Fog* (San Francisco: Sierra Club Books, 1983), pp. 1–34, 199–241.

[51] E. G. Vallianatos with Mckay Jenkins, *Poison Spring* (London: Bloomsbury Publishing, 2014), pp. 103–120.

[52] *Ibid.*, pp. 23–236.

Reform or Ban Industrial Agriculture and Animal Farms

The second source of the pandemic and climate chaos has always been giant and toxic agribusiness.[53] The US Centers for Disease Control and Prevention (CDC)[54,55] have warned us about the pathogenic zoonotic diseases emerging in the animal feeding enclosures and from the animals themselves. COVID-19 is a zoonotic infectious disease. About 60% of all infectious disease agents originate in animals. Some 80% of diseases with bioterrorism potential are zoonotic diseases. To industrial agriculture and animal farms, add fur farming and pet trade industry, and the risk to Americans is "significant [and] staggering." Animal factories could trigger "infectious disease outbreaks."[56] They did seemingly give us the pandemic of 2020–2022. Animal farms or Concentrated Animal Feeding Operations (CAFOs), says Helena Masiello,[57] Professor of Law at the University of Miami, "are a breeding ground for deadly infectious diseases."

I already explained above why the eating of meat is antithetical to our aspirations for fighting climate change. Of course, some people have legitimate reasons for eating meat. But, in general, it would be a good idea to stop eating animals. That way you throw a huge wrench in the gears of factory plagues. Stop being a consumer of other living creatures. In my personal view, humans do not need to eat meat. This simple act will tell those selling meat as well as political lobbyists that you are not going to continue supporting their hazardous business. Abandoning meat could potentially help our chances of surviving the colossal climate chaos already in our backyard.

---

[53] Andrew Kimbrell, ed., *Fatal Harvest: The Tragedy of Industrial Agriculture* (Washington, DC: Island Press, 2002).

[54] Carrie Hribar, Understanding concentrated animal feeding operations and their impact on communities, CDC, 2010.

[55] R. Ghai and C. Barton Behravesh, Zoonoses – The one health approach, In *CDC Yellow Book 2024*: Environmental Hazards & Risks.

[56] Emily Anthes, The risk is staggering, report says of disease from U.S. animal industries, *New York Times*, July 7, 2023.

[57] Helena Masiello, CAFOs are a public health crisis: The creation of COVID-19, *University of Miami Law Review*, 76 (3), June 7, 2022.

In 2023, the US Conference of Mayors[58] took a step in that direction. It adopted a resolution preferring green plants to eating meat. "The United States Conference of Mayors," the resolution read, "supports [eating fruits and vegetables in order] to combat chronic disease [and] the climate crisis… [The US Conference of Mayors is] exploring opportunities to include more plant-based options in any setting where city government provides food to constituents (schools, hospitals, social services), exploring opportunities to promote the benefits of a plant-based approach to constituents through public awareness campaigns, exploring opportunities to evaluate the environmental impact of food choices and move toward a more plant-centered approach for individual and population health, as well as local and global environmental wellbeing." If the plant-based foods are made by a variety of plants/crops grown without synthetic fertilizers and pesticides, then plant-based foods are reasonable alternatives to eating meat.

Brewing Pandemics

I already said that the Greek philosopher, mathematician, and cosmologist, Pythagoras, praised the beauty, perfection, and harmony of the natural world and the Cosmos. He urged people never to kill, much less eat, another living being.[59] The insights of Pythagoras on animals and vegetarianism were of enormous and lasting importance. It was said that they influenced Plato and Aristotle, who invented the science of zoology. I cited Aristotle in the beginning of this Conclusion. He praised the perfection of nature and animals in particular. Animals, including insects, keep ecosystems functioning and alive. They give us a healthy planet. Can you imagine nature without birds, honeybees, Monarch butterflies, dogs, cats, donkeys, mules, horses, sheep, goat, cattle, lions, primates, elephants, dolphins, fish, whales, wolves, or eagles? I cannot. Humans must lift their heads from the sand. They need to (re)read Aristotle's *History of Animals* and *Parts of Animals*. The great philosopher and inventor of the

---

[58] 1400 Mayors Join New York Mayor Eric Adams in promoting plant-based food, *VegNews*, July 31, 2023. https://vegnews.com/1400-mayors-plant-based-food.
[59] Iamblichus, *On the Pythagorean Life*, pp. 6–7, 99–100.

field of formal logic[60] said animals are works of nature. They are beautiful and perfect.[61] As David Benatar, Emeritus Professor of the University of Cape Town, South Africa said in 2007 in the *American Journal of Public Health,* humans must recognize "the risk to themselves that can arise from their maltreatment of other species."[62]

In domestic affairs, especially those of food and agriculture and public health, we should be honest. The *Scientific American Magazine*[63] warns that we need to admit we are "brewing" pandemics in our backyards "through our risky animal-use practices." It goes on to say that we should know "that this [2020] catastrophe may be just a dress rehearsal for an even more serious pandemic that could take a more gruesome toll – akin to the 1918 global flu pandemic… When that day comes, it's very likely that such a virus will also have its origin in humanity's seemingly insatiable desire to eat animals, whether wild or domestic."

Another thinker and contributor to the *New York Times Magazine,* Ferris Jabr,[64] explained our predicament in terms of what we are doing to disrupt the microscopic world of the invisible inhabitants of the planet, fungi, microbes and viruses. "More than any other entity," he said, "viruses and microorganisms expose the fallacy of our tyrannical choreography. We are wed to thinking of ourselves as the protagonists of every landscape, but from the perspective of infectious microbes, we and other large creatures are the landscape. As we restructure Earth's biosphere to suit our whims, we open hidden conduits between other animals' microbiomes and our own. Once those channels are in place, pathogens can no more stop themselves from spilling into us than water can prevent itself from running downhill. We cannot blame the bats, mosquitoes and viruses. We cannot expect them to go against their nature. The challenge before us is how best to govern ourselves and stymie the flood we unleashed."

---

[60] Aristotle (2024), *Encyclopædia Britannica.* https://www.britannica.com/biography/Aristotle (Accessed: 27 September 2024).

[61] Aristotle, *Parts of Animals,* 644b28–645a7-30.

[62] David Benatar, The chickens come home to roost, *American Journal of Public Health,* September 2007.

[63] Paul Shapiro, One root cause of pandemics few people think about, *Scientific American,* March 24, 2020.

[64] Ferris Jabr, How humanity unleashed a flood of new diseases, *New York Times Magazine,* June 25, 2020.

Human desires for becoming the architects of an anthropomorphic and anthropocentric planet are no less powerful than those desires to keep eating animal flesh, showing the non-human denizens who the bosses are. Meat eating continues unabated. The meat we eat comes from the animals concentrated in the thousands of animal farms across the globe. Unfortunately, as I have demonstrated in Chapter 6 and 7, these animal farms are far from clean and can potentially be the cause of viral pandemics. I repeat: the easiest thing we can do at this time is to stop eating animals. "A growing number of Americans," after all, as noted by the Animal Welfare Institution in 2020,[65] have already begun "celebrating the [winter] season with cruelty-free, plant-based protein options."

I believe that a safe alternative is organic family farming, which grows food without toxic fertilizers and pesticides. It's also important that organic farming is family-owned; not corporate in ambition, reality, and effects. Small-scale organic family farming can feed the nation and revitalize rural America. It can mitigate the effects of climate change. In my opinion, it is also science-based and theoretically pandemic-free agriculture that is very productive and respectful of domesticated animals and wildlife.

Climate chaos, however, threatens not only industrialized agriculture, but also small-scale agroecological farming. Heavy rains and lasting draughts harm both. Professor of Agroecology at the University of California-Berkeley, Miguel Altieri, is studying the effects of climate change on all kinds of farms. He observed that the resilience of family farms is not unlimited. There are limits to socioecological resilience, Altieri says.[66] Something must be done to slow down the burning of fossil fuels which keeps strengthening climate chaos. In other words, Altieri says, we must do more than raising organic food. Organic farming is fundamental but climate is in the room. Climate will change everything we take for granted.

---

[65] Animal Welfare Institute (2020), *Op. cit.*
[66] Miguel A. Altieri, Agroecology and Climate Change: The Limits of Socioecological Resilience, *Revista FORO*, November/December 2023, pp. 1–7.

Cooperate or Perish!

In Chapter 6, I cited UN chief, Antonio Guterres,[67] warning prime ministers and presidents that climate danger is all too real, and a threat to humanity and the planet. He told them: "cooperate or perish!"; urging them to form a climate solidarity pact. Failing that, he told them, they would end up forming a collective suicide pact. Guterres reminded and warned world leaders that humanity is already on the precipice of no return.

A Moment of Climate Truth

On June 5, 2024, Guterres forcefully repeated his warnings.[68] This is a moment of truth, he said to his audience at the Natural History Museum of New York. He said people the world over are in danger from climate chaos. Yet he reminded them that people themselves were the danger. Their activities, especially the burning of fossil fuels, threatened the planetary temperatures higher than 1.5°C above pre-industrial times. "The difference between 1.5 and two degrees," Guterres explained, "could be the difference between extinction and survival for some small island states and coastal communities. The difference between minimizing climate chaos or crossing dangerous tipping points. 1.5 degrees is not a target. It is not a goal. It is a physical limit. Scientists have alerted us that temperatures rising higher would likely mean: The collapse of the Greenland Ice Sheet and the West Antarctic Ice Sheet with catastrophic sea level rise; The destruction of tropical coral reef systems and the livelihoods of 300 million people; The collapse of the Labrador Sea Current that would further disrupt weather patterns in Europe; And widespread permafrost melt that would release devastating levels of methane, one of the most potent heat-trapping gasses. Even today, we're pushing planetary boundaries to the brink – shattering global temperature records and reaping the whirlwind. And it is a travesty of climate justice that those least responsible for the crisis are hardest hit: the poorest people; the most vulnerable countries; Indigenous Peoples; women and girls. The richest one per cent emit as much as two-thirds of humanity."

---

[67]National Public Radio, The U.N. chief tells the climate summit cooperate or perish. Sharm el-shekh, Egypt, November 7, 2022.

[68]Antonio Guterres, A moment of truth, *UN News*, June 5, 2024.

So, Guterres, once again, warned policy makers what is at stake. To achieve that 1.5°C (and no higher), about 50% of fossil fuels would have to be phased out by 2030. But, once again, it is unlikely that the industry and politicians took him seriously. Industry lobbyists routinely deny the danger of burning fossil fuels. They fund academics to raise doubts about climate change.[69] Such measures of deception have precedents. Take tobacco companies[70] as an example. They kept their cancer-causing cigarettes on the market for more than a century. They worked together with Hollywood to make cigarettes desirable objects. Tobacco companies convinced Hollywood and the film industry to embrace cigarettes. Most movies done in the twentieth century are filled with people smoking and drinking. While no longer as prevalent, tobacco products still make appearances on the silver screen, and cigarettes still sell today, though at reduced rates.

The chemical industry, and especially the pesticide companies, learned the tobacco lesson too well. For several decades, farmers have been spraying their crops with neurotoxic and carcinogenic pesticides.[71] They listened to the advertisements of the USDA and received their spray license and registration from the EPA. They convinced themselves that those toxins were safe. While the EPA banned dichloro-diphenyl-trichloroethane (DDT) and a few other DDT-like chemicals in the 1970s, they later froze and allowed the pesticide companies to set its regulatory agenda both under Republican and Democratic administrations. I was there and observed the changes.[72]

The fossil fuel industry learned and taught deception in hiding the dangers of petroleum, coal, and natural gas. Exxon scientists, for example, warned the company's executives of the climate change effects of petroleum burning in millions of cars, factories, airplanes, and other machines as far back as in the 1970s.[73] But the Exxon (and other fossil fuel industry)

---

[69] Naomi Oreskes and Erik M. Conway, *Merchants of Doubt* (New York: Bloomsbury Press, 2010), pp. 169–215.

[70] National Library of Medicine: C. Mekemson and S. Glantz, How the tobacco industry built its relationship with Hollywood, *Tobacco Control*, March 2002.

[71] Robert van den Bosch, *The Pesticide Conspiracy* (Berkeley, CA: University of California Press, 1989), pp. 57–79.

[72] E. G. Vallianatos with Mckay Jenkins, *Poison Spring* (London: Bloomsbury Publishing, 2014), pp. 23–230.

[73] Shannon Hall, Exxon Knew about Climate Change about 40 years ago, *Scientific American*, October 26, 2015.

executives dismissed the findings of their own scientists. They started funding university scientists to produce doubts about climatological science and confuse politicians and the American people. They bought science in favor of continuing profits. Doubt, Oreskes and Conway, said became their chief product.[74] They continue this deception to this very day.[75] The Republicans do the same thing. They continue to deny climate change.[76] They rather risk America, their own children, and the world than lower their petroleum, natural gas, and coal profits. They have even organized the attorneys general in Republican/red states to silence the EPA on fossil fuel regulations – and more. Republicans employ "conservative legal activists and their funders" tied to the coal and oil industries in order to force the rewriting of environmental laws and regulations, thus weakening the "executive branch's ability to tackle global warming."[77] But bad policies come from the Democrats as well. In mid-December 2022, California's Public Utilities Commission influenced Gavin Newsom, governor of California, to slash the incentives of putting solar panels on house rooftops.[78]

Retribution

The governor of California is not a stupid man. During the frequent fires consuming the forests of the state, he said, yes, climate change is the real enemy. However, he diminished the subsidies for the use of solar panels in California as cited above. He also employed litigation against the largest oil companies.[79] The State of California sued Exxon, Shell, BP, Chevron, and ConocoPhillips for deception. The allegation proposed that these giant companies were fully aware of the climate consequences of the burning of their products – yet they misled Americans and their leaders for decades. The suit accused companies of professing that there was no

[74]Naomi Oreskes and Erik M. Conway, *Merchants of Doubt* (New York: Bloomsbury Press, 2010).

[75]Evaggelos Vallianatos, Big oil and civilization don't mix, *Counterpunch*, May 17, 2024.

[76]Lisa Friedman and Jonathan Weisman, Delay as the new denial: The latest republican tactic to block climate action, *New York Times*, July 20, 2022.

[77]Coral Davenport, Republican drive to tilt courts against climate action reaches a crucial moment, *New York Times*, June 28, 2022.

[78]Sammy Roth, California just slashed rooftop solar incentives. What happens next? *Los Angeles Times*, December 15, 2022.

[79]David Gelles, California sues giant oil companies, citing decades of deception, *New York Times*, September 15, 2023.

connection between fossil fuels and storms, heat waves, floods, tornadoes, wildfires, crop damage, and the decimation of biological diversity. Or as stated in the California legal complaint charges:

> *Oil and gas company executives have known for decades that reliance on fossil fuels would cause these catastrophic results, but they suppressed that information from the public and policymakers by actively pushing out disinformation on the topic... Their deception caused a delayed societal response to global warming. And their misconduct has resulted in tremendous costs to people, property, and natural resources, which continue to unfold each day. This has been a multi-decade, ongoing campaign to seek endless profits at the expense of our planet, [and] our people.*

Rob Bonta, the attorney general of California, made these charges. He also said that the greedy corporations and individuals need to be held accountable... That's where we come in. Gavin Newsom expressed similar sentiments. He was astonished with the violence of climate change. 'This last 10 years, it's shook me to my core,' he said. "These are things that we imagined we might be experiencing in 2040 and 2050, but that have been brought into the present moment, and the time for accountability is now." By accountability, Newsom and Bonta do not mean anything like putting these unethical and dangerous companies out of business. No. Bonta expects these giant oil and gas companies to cough up large amounts of money to expiate their crimes. California will then employ these funds to pay for recovery from extreme weather events, mitigation and adaptation efforts across the state.

The legislature of California also passed a number of laws that lessen to some degree the fossil fuel emergency. For instance, companies earning more than US$ 1 billion per year will be required to disclose their carbon footprint; large oil and gas companies will not be allowed to pass off their old oil wells to smaller companies unable to clean up those wells; the state will make it difficult to drill for oil and gas along California's Pacific coast; the state of California will buy electricity from offshore wind farms; cities will be able to prohibit plastic grass; drinking water can no longer be used for purely decorative grass; and school buses are required to be electric by 2030 – to name a few.[80]

---

[80] Sammy Roth, Here are all the climate and environmental bills that California just passed, *Los Angeles Times*, September 19, 2023.

These policies, especially the ones seeking damages from oil and gas companies, are potentially good and useful, but they are of limited value. It may take decades of litigation by which time California may no longer be California. The year 2030 by which we need to end the use of at least 43% of all fossil fuels is fast approaching. But judging from the climate policies of California, the United States, and the rest of the oil-captured world, it appears to me that 2030 is likely to be another normalized year of burning fossil fuels, resulting in catastrophic consequences for the Earth and civilization. While Guterres has repeatedly issued his dire warnings, prime ministers and presidents have continued to ignore them and Guterres. I suspect that like most people, world leaders probably think that climate change, if it exists at all, is like any other problem of a developed world. But it is not. Global warming is like an exploding nuclear bomb that affects the whole world. Just as the only way to eliminate nuclear war is to eliminate nuclear weapons; similarly, to eliminate the destruction that the continual burning of fossil fuels cause requires the elimination of the consumption of fossil fuels – and as soon as possible. Merely wishing for the climate emergency to go away is not enough.

Newsom and California seems to be ambivalent about the climate emergency based on their actions. Californians love their private cars. Yes, Newsom is suing the very companies causing global warming, but he appears to be still listening to the enemies of the present and the future – fossil fuel executives, and utility men hooked on burning petroleum, natural gas, and coal for electricity production. Putting a brake on the adoption of solar panels was decisively negative. This action sent a message to Big Oil that they do not need to worry. The mixed messages are unlikely to reduce the burning of fossil fuels by 2030.

Environmentalists: Defend America and the World

One of the Aesopian Myths[81] says: "Pray to Athena, but do something about the favor you are seeking from the goddess" – "Σύν Αθηνά και χεῖρα κίνει." In the same spirit today, we know that climate change is here to stay. Our duty is to do something about the crisis and, simultaneously,

---

[81]*Aesop's Myths: The Shipwrecked Rich Athenian and Athena.* There's an excellent English translation of the myths of Aesop: Laura Gibbs, tr., *Aesop's Fables* (New York: Oxford University Press, 2002).

to pray to Athena, which in our time would mean to talk to our politicians to face the facts and fight climate chaos. The time has come for Americans – especially climate scientists who know what the dangers of global warming are – and all environmentalists and their organizations, to come together and fight to mitigate the giant climate danger hovering over us all. We are today, seemingly where Americans in the 1980s were; powerlessly watching the threat of an imminent nuclear confrontation between America with the Soviet Union looming on the horizon. Scientists then predicted an uninhabitable Earth and permanent winter should the Cold War became a nuclear war. Fortunately, Presidents Ronald Reagan and Mikhail Gorbachev were frightened enough that the Cold War remained cold. Now, some 40 years later, the Second Cold War is dangerously heating up. And strangely enough, we (America, Russia, and the rest of the world) are facing the angry whiplash of climate chaos, entirely anthropogenic like the nuclear weapons crisis. It is as if humans are trying to bring another asteroid against the planet. A similar stone mass wiped out the dinosaurs 66 million years ago.[82] Do the "leaders" of humanity want another asteroid to hit the planet? Or, allow the consequences of catastrophic climate chaos precipitate a nuclear war? Such a prospect should be inconceivable. In 1963, President John Kennedy said "a single nuclear weapon contains almost ten times the explosive force delivered by all the allied air forces in the second World War."[83]

Those of us who clearly see our predicament need to talk to each other; having national conferences to warn others of the climate threat is clearly not enough. We probably would need to take more actions, perhaps organizing a ground-up initiative to transition toward a new way to life without fossil fuels. While the super-rich could possibly advance the transition to non-carbon technologies, the unfortunate reality is that we have not progressed very far from where we were in the 1960s, toward our goal of being weaned off fossil fuels; as seen in an interactive Global fossil fuel chart on Our World in Data[84]; in fact, the quantity of fossil fuels

---

[82] Emily Osterioff, How an asteroid ended the age of the dinosaurs, Natural History Museum. https://www.nhm.ac.uk/discover/how-an-asteroid-caused-extinction-of-dinosaurs.html.

[83] President John F. Kennedy, Commencement Address at American University, Washington, D.C., June 10, 1963. https://www.jfklibrary.org/archives/other-resources/john-f-kennedy-speeches/american-university-19630610.

[84] Hannah Ritchie and Pablo Rosado (2017), Fossil fuels. Published online at OurWorldinData. org. Retrieved from https://ourworldindata.org/fossil-fuels [Online Resource].

consumed has only gone up (with the sole significant dip being during the global lockdown). I have repeatedly mentioned that the UN Chief, Guterres, had said that keeping the global temperature at 1.5°C above preindustrial levels requires that we shut down half of all fossil fuel consumption during this decade. So, in order for enough actions to be taken, environmentalists need to explain how they envision the future in which the current system focused on fossil fuels is replaced with an alternative system and with people committed to green technologies capable of achieving net-zero emissions.

We can also improve what we have started in our plans of phasing out of fossil fuels. The adoption of electric cars is just the beginning. We can do more. Sadly, I believe that ill will, dysfunction, insults and shoutdowns in Congress, the wars in Ukraine and Israel–Palestine, as well as a broken Capitol will keep fossil fuel companies in business. We saw the consequences of an unstable democracy in the January 6, 2021 Insurrection. The fallout of the failed coup is a permanent reminder that democracy is not secure in America. At the end of 2021, there was "a sense that Congress was not rising to meet a perilous moment in history."[85] Trump then declared his candidacy for the 2024 elections. In fact, he was more than the Republican candidate. He won the 2024 election and has been appointed the President of the country for the second time. Karen Greenberg, professor of law at Fordham University, sees the abuse of power by Trump like a mirror of corruption and abuse of power: "the abuse of power at the absolute highest level of government to harm the country."[86] Yes, indeed, to harm the country. But the Supreme Court intervened and gave much more power to the President, including President Trump.

Biden's Opportunity and Failure

The climate funding President Biden signed into law during his term was a measure and inspiration to do some of what was necessary to save the country from climate calamity. President Biden, however, did not fulfill his election promises. He said he would no longer allow the oil and gas

[85] Jonathan Weisman, Congress ends "horrible year" with divisions as bitter as ever, *New York Times*, December 18, 2021.
[86] Democracy Now!, August 15, 2022.

companies to lease public land for petroleum and gas development. He did the exact opposite. On October 7, 2021, American research scientists and climatologists sent Biden a letter[87] in which they criticized him for opening public lands and ocean waters for oil and gas extraction. Biden's actions infuriated some environmentalists so much, they referred to his opening public lands for oil extraction as "Biden's climate bomb."[88] According to Gladys Delgadillo of the Climate Law Institute and Brady Bradshaw of the Center for Biological Diversity: "The simple, uncompromising truth is that climate change will affect all of us, and new federal fossil fuel leasing is a death sentence for people and wildlife on the frontlines of the climate emergency. If... president [Biden] won't stop fossil fuel investment on the lands he directly controls, what hope is there that he can encourage the rest of the nation and [the] world to end the fossil fuel era?"[89] Exactly. In fact, Biden went further in disregarding his climate promises. On March 13, 2023, he approved the operation of the Willow Project in the Arctic of Alaska. The implications, ecological and human, of oil extraction in the country's most beautiful and unspoiled 23 million acres of land are staggering. Greenhouse gas emissions will increase; the disruption and pollution caused by road building, and the noise from mining the land for petroleum will harm ecosystems, wildlife, and the indigenous people.[90]

As I already said, I firmly believe that America needs to stop the wars in Ukraine and Israel. America needs to take the first step to invite President Vladimir Putin of Russia, President Xi Jinping of China, Prime Minister of India Narendra Modi, heads of state of the European Union, and other leaders for an emergency climate summit to coordinate the transition to green energy. In my opinion, the ideal outcome of such a summit would be for the heads of states to sign an Earth-preserving treaty in which nation states become responsible for cutting greenhouse gases by about 50% well before 2030. In my ideal world, if politicians refuse to touch the fossil-fuel companies and pollution, then the environmentalists

---

[87] An Open Letter from US Scientists Imploring President Biden to End the Fossil Fuel Era, October 7, 2021. Science and Environmental Health Network.

[88] Basay Sen, Biden's dangerous climate hypocricy, *The Progressive Magazine*, August 15, 2023.

[89] Center for Biological Diversity, *Endangered Earth*, Summer 2022.

[90] Evaggelos Vallianatos, Willow project: A carbon bomb is exploding in the Arctic of Alaska, *Counterpunch*, March 16, 2023.

must run for elections to ensure that state power is supportive in guiding the world into a green future and civilization.[91]

This climate emergency should be in the mind and hearts of all thinking human beings. Yes, the wars currently occurring are tragedies that must end with negotiations, but they should not divert public and international attention from the prize: the importance of a better future and reversing the destruction of the planet.

The UN Chief, Antonio Guterres warned the world leaders, once again on September 20, 2023: He said, "Humanity has opened the gates of hell. Horrendous heat is having horrendous effects. Distraught farmers watching crops carried away by floods. Sweltering temperatures spawning disease. And thousands fleeing in fear as historic fires rage. Climate action is dwarfed by the scale of the challenge. If nothing changes, we are heading towards a 2.8°C temperature rise – towards a dangerous and unstable world."[92]

Less than a month later, on October 4, 2023, Pope Francis[93] published a letter on climate change. The long letter was based on an admirable knowledge of climate science. The late Pope was angry with the rulers of the world for risking life on Earth. "I have realized," he said, "that our responses [to climate crisis] have not been adequate, while the world in which we live is collapsing and may be nearing the breaking point." I agree, the world is collapsing precisely because its leaders have not done enough to mitigate the damage from fossil fuels. Five weeks after the warning of the Pope, on November 14, 2023, the US government issued the *Fifth National Climate Assessment*,[94] which warned Americans that their country was warming much faster than any other country, a fact explaining why climate change was unsettling their lives. Then in December 2023 the Climate Summit took place in the kingdom of the United Arab Emirates.[95] Apparently, the autocratic leaders of UAE realize

---

[91] Evaggelos Vallianatos, Climate nemesis: Hurtling toward a universal world, *Counterpunch*, July 3, 2022.

[92] UN Secretary General, Secretary-general opening remarks at the climate ambition summit, UN, New York, September 20, 2023.

[93] Pope Francis, *Laudate Deum/Praise God*, October 4, 2023.

[94] US Fifth National Climate Assessment, November 14, 2023. https://nca2023.global change.gov.

[95] UN Climate Change Conference – United Arab Emirates, November 30 – December 12, 2023. https://unfccc.int/cop28.

Fig. 1. Tree of Life, Prayer for the Planet. Painting by Katherine Ferwerda, graduate student, Fall 2021, in my seminar (regeneration: concepts and applications) at the Lyle Center for Regenerative Studies, California State Polytechnic University, Pomona. Courtesy of Katherine Ferwerda.

that climate change is serious, and certain to affect their country in extremely adverse ways. They try to address those "urgent dilemmas." Scientists probably told them and the world that the goal of keeping the temperature of the planet not higher that 1.5°C is almost a lost battle. It is "out of reach." The temperature of the planet rose by 1.2°C, not that far from the 1.5°C, the point above which humans would find it difficult to face the rising heat waves, thunderous storms, hurricanes, forest and cities fires, and drought.[96]

I urge all Americans, the people of the planet, and environmentalists in particular to raise the flag of virtue and defend our civilization and Mother Earth.

In that process of regeneration, moving away from fossil fuels and embracing net-zero carbon-emitting technologies of solar and wind energy, we must return to sacred and democratic family farming and other core values of Greek and Western civilization – namely, democracy, justice, compassion and love for animals, love for Mother Earth, freedom of speech, and the good and the beautiful.

---

[96] Vivian Nereim, The dilemmas of a petrostate preparing to host a climate summit, *New York Times*, November 29, 2023.

Acknowledgments

I thank Michael Fox and Miguel Altieri for their life-long struggles to improve life on Earth. But above all I thank small family farmers for preserving and practicing ancient agroecological and democratic farming. I am also very grateful to my editors, Amanda Yun and Zerlina Zhuang, for their insights that made this a better book.

Bibliographic Note

Footnotes show my sources, most of which are newspapers and reports. Readers interested in following up on the arguments of the book can track down those articles and reports and the few books I cite. The issues of environmental protection, animal farms, industrialized farming, pesticides, biological warfare laboratories, and the pandemic are extremely controversial. My suggestion is that experts may not necessarily reveal everything that they know, if they publish at all.

I do not intend to reproduce the footnotes in the bibliography, which will be very select. The books in the bibliography present broad studies examining historical, agricultural, ecological, scientific, and political views. They help us understand the politics behind the destruction of the natural world. Some of them explain the reasons why a substantial fragment of democratic and productive family farming disappeared or became factories in the field.

Bibliography

Allen, W. *The War on Bugs*. White River Junction, VT: Chelsea Green Publishing, 2008.

Barlow, M. *Blue Future: Protecting Water for People and the Planet Forever*. New York: The New Press, 2014.

Bergland, B. *A Time to Choose: Summary Report on the Structure of Agriculture*. Washington, DC: U.S. Department of Agriculture, 1981.

Borkin, J. *The Crime and Punishment of I. G. Farben*. New York: The Free Press, 1978.

Bosch, R. Van Den. *Pesticide Conspiracy*. Berkeley, CA: California University Press, 1989.

Carson, R. *Silent Spring*. Boston: Houghton Mifflin Company, 1962.

Conis, E. *How to Sell a Poison: The Rise, Fall, and Toxic Return of DDT*. New York: Bold Type Books, 2022.

Crosby, A. W. *Ecological Imperialism: The Biological Expansion of Europe, 900–1900*. New York: Cambridge University Press, 1990.

Devall, B. (ed.). *Clearcut: The Tragedy of Industrial Forestry*. San Francisco: Sierra Club Books/Earth Island Press, 1993.

Drucker, S. M. *Altered Genes, Twisted Truth: How the Venture to Genetically Engineer Our Food Has Subverted Science, Corrupted Government, and Systematically Deceived the Public*. Salt Lake City, UT: Clear River Press, 2015.

Epstein, S. S. *Cancer-Gate: How to Win the Losing Cancer War*. Amityville, NY: Baywood Publishing Company, 2005.

Fowler, C. and Mooney, P. *Shattering: Food, Politics, and the Loss of Genetic Diversity*. Tucson: University of Arizona Press, 1990.

Gross, P. R., Levitt, N. and Lewis, M. W. (eds.). *The Flight From Science and Reason*. New York: The New York Academy of Sciences, 1996.

Hansen, J. *Storms of My Grandchildren: The Truth About the Coming Climate Catastrophe and Our Last Chance to Save Humanity*. New York: Bloomsbury, 2010.

Hauter, W. *Frackopoly: The Battle for the Future of Energy and the Environment.* New York: The New Press, 2016.

Helias, P.-J. *The Horse of Pride: Life in a Breton Village.* Tr. J. Guicharnaud. New Haven: Yale University Press, 1978.

Hesiod. *Theogony, Works and Days, Shield.* Tr. A. N. Athanassakis. Baltimore, MD: Johns Hopkins University Press, 1983.

Jackson, W. *New Roots for Agriculture.* Lincoln, Nebraska: University of Nebraska Press, 1980.

Jackson, W. and Jensen, R. *An Inconvenient Apocalypse: Environmental Collapse, Climate Crisis, and the Fate of Humanity.* Notre Dame, IN: University of Notre Dame Press, 2022.

Ketcham, C. *This Land: How Cowboys, Capitalism, and Corruption are Ruining the American West.* New York: Viking, 2019.

Kimbrell, A. (ed.). *Fatal Harvest: The Tragedy of Industrial Agriculture.* Washington, DC: Island Press, 2002.

Kolbert, E. *Field Notes from a Catastrophe: Man, Nature, and Climate Change.* New York: Bloomsbury, 2006.

Krebs, A. V. *The Corporate Reapers: The Book of Agribusiness.* Washington, DC: Essential Books, 1992.

Leroi, A. M. *The Lagoon: How Aristotle Invented Science.* New York: Viking, 2014.

Markowitz, G. and Rosner, D. *Deceit and Denial: The Deadly Politics of Industrial Pollution.* Berkeley: University of California Press, 2002.

Marsh, G. P. *Man and Nature Or, Physical Geography as Modified by Human Action.* Ed. D. Lowenthal. Cambridge, MA: Harvard University Press, 1967.

McCully, P. *Silenced Rivers: The Ecology and Politics of Large Dams.* London: Zed Books, 1996.

McPhee, J. *The Control of Nature.* New York: Farrar, Straus, and Giroux, 1989.

Mitchell, J. *Poisoning the Pacific: The US Military's Secret Dumping of Plutonium, Chemical Weapons, and Agent Orange.* Lanham, MD: Rowman and Littlefield, 2020.

Mooney, C. *The Republican War on Science.* New York: Basic Books, 2005.

Oreskes, N. and Conway, E. M. *Merchants of Doubt: How a Handful of Scientists Obscured the Truth on Issues from Tobacco Smoke to Global Warming.* New York: Bloomsbury Press, 2010.

Osborne, R. *Classical Landscape With Figures: The Ancient Greek City and its Countryside.* Dobbs Ferry, NY: Sheridan House, 1987.

Porritt, J. (ed.). *Save the Earth.* Atlanta, GA: Turner Publishing, 1991.

Reisner, M. *Cadillac Desert: The American West and its Disappearing Water.* New York: Penguin Books, 1987.

Rich, B. *Mortgaging the Earth: The World Bank, Environmental Impoverishment, And the Crisis of Development.* Boston: Beacon Press, 1994.

Robin, M.-M. *Our Daily Poison: From Pesticides to Packaging, How Chemicals Have Contaminated the Food Chain and Are Making Us Sick.* Tr. A. Schein and L. Vergnaud. New York: The New Press, 2014.

Rosenblum, M. *Olives: The Life and Lore of a Noble Fruit.* New York: North Point Press, 1996.

Russo, L. *The Forgotten Revolution: How Science Was Born in 300 BC and Why It Had to Be Reborn.* Tr. S. Levy. Berlin: Springer, 2004.

Schlosser, E. *Fast Food Nation: The Dark Side of the All-American Meal.* Boston: Houghton Mifflin Company, 2001.

Schweiger, L. J. *Climate Crisis and Corrupt Politics: Overcoming the Powerful Forces that Threaten Our Future.* Irvine, CA: Universal Publishers, 2019.

Sherman, J. D. *Life's Delicate Balance: Causes and Prevention of Breast Cancer.* New York: Taylor and Francis, 2000.

Silbergeld, E. *Chickenizing Farms and Food: How Industrial Meat Production Endangers Workers, Animals, and Consumers.* Baltimore: Johns Hopkins University Press, 2016.

Speth, J. G. *Red Sky at Morning: America and the Crisis of the Global Environment.* New Haven: Yale University Press, 2004.

Stoyannis, V. and P. Dilana (eds.). *The Odyssey of the Greek Agricultural Biodiversity.* Athens: Odyssey Network/Nea Ecologia, 2001.

Strum, C. Van. *A Bitter Fog: Herbicides and Human Rights.* San Francisco: Sierra Club Books, 1981.

Torrance, R. M. (ed.). *Encompassing Nature: A Sourcebook: Nature and Culture from Ancient Times to the Modern World.* Washington, DC: Counterpoint, 1999.

Vallianatos, E. *Fear in the Countryside: The Control of Agricultural Resources in the Poor Countries by Non-Peasant Elites.* Cambridge, MA: Ballinger Publishing Company, 1976.

Vallianatos, E. *Harvest of Devastation: The Industrialization of Agriculture and its Human and Environmental Consequences.* Goa, India: The Other India Press/New York: The Apex Press, 1994.

Vallianatos, E. *The Antikythera Mechanism: The Story Behind the Genius of the Greek Computer.* Irvine, CA: Universal Publishers, 2021.

Vallianatos, E. *This Land is Their Land: How Corporate Farms Threaten the World.* Monroe, ME: Common Courage Press, 2006.

Vallianatos, E. and Jenkins, M. *Poison Spring: The Secret History of Pollution and the EPA.* New York: Bloomsbury Press, 2014.

Wallace, D. R. *Life in the Balance.* New York: Harcourt Brace Jovanovich Publishers, 1987.

White, K. D. *Country Life in Classical Times.* Ithaca, NY: Cornell University Press, 1977.

Wilcox, F. A. *Scorched Earth: Legacies of Chemical Warfare in Vietnam*. New York: Seven Stories Press, 2011.

Worster, D. *Rivers of Empire: Water, Aridity, and the Growth of the American West*. New York: Oxford University Press, 1985.

Xenophon. *Oeconomicus: A Social and Historical Commentary*. Tr. S. B. Pomeroy. New York: Oxford University Press, 1994.